FOREST AND LABOR IN MADAGASCAR

FOREST AND LABOR IN MADAGASCAR

FROM COLONIAL CONCESSION TO GLOBAL BIOSPHERE

GENESE MARIE SODIKOFF

Indiana University Press
Bloomington & Indianapolis

This book is a publication of

Indiana University Press
601 North Morton Street
Bloomington, Indiana 47404-3797 USA

iupress.indiana.edu

Telephone orders 800-842-6796
Fax orders 812-855-7931

∞ The paper used in this publication meets the minimum requirements of the American National Standard for Information Sciences—Permanence of Paper for Printed Library Materials, ANSI Z39.48-1992.

Manufactured in the United States of America

Library of Congress Cataloging-in-Publication Data

Sodikoff, Genese Marie, [date]
Forest and labor in Madagascar : from colonial concession to global biosphere / Genese Marie Sodikoff.
p. cm.
Includes bibliographical references and index.
ISBN 978-0-253-00309-6 (cloth : alk. paper) — ISBN 978-0-253-00577-9 (pbk. : alk. paper) — ISBN 978-0-253-00584-7 (electronic book) 1. Forest conservation—Madagascar. 2. Forest biodiversity conservation—Madagascar. 3. Forests and forestry—Economic aspects—Madagascar. 4. Rural poor—Madagascar. 5. Madagascar—Economic conditions. I. Title.
SD414.M28S63 2012
333.75'1609691—dc23

2012010378

1 2 3 4 5 17 16 15 14 13 12

TO MY PARENTS, GARY AND EMILY,
for all their worries and care,

TO HANK,
for so much support,

TO OSCAR, VINNIE, AND SCARLETT,
my little dreams come true,

AND IN MEMORY OF MY MOTHER, INEZ,
with love

Arbres sur la colline où reposent nos morts
dont l'histoire n'est plus, pour ma race oublieuse,
que fable, et toi, vent né des zones soleilleuses
qui ranimes leur sein d'ombre humide et le mords,

ce soir, je vous contemple et mon coeur vous écoute:
votre rumeur me dit l'âme de mes aïeux
tandis que l'horizon tragique et radieux
    annonce d'un beau jour la gloire et la déroute.

Trees on the hill where our dead rest,
whose history, for my forgetful race, is now
but fable, and you, wind born of sunny zones,
who revive and bite their chest of moist shadow,

tonight, I ponder you and my heart listens:
your sound recounts the soul of my forefathers
while the tragic and radiant horizon
    announces with dawn glory and ruin.

—JEAN-JOSEPH RABEARIVELO, *Volumes*
(translated from the French by Richard Serrano)

CONTENTS

ACKNOWLEDGMENTS

This is an account of the evolving sensibilities of time, space, and nature in Madagascar that take shape through obdurate social hierarchies and ethnocentrisms, as well as through tactile encounters with land and wildlife. As a history of deforestation and underdevelopment, the book examines the entwinement of these processes and reflects my view that social structures based on degrading practices and belief systems—degrading of persons, societies, and ecologies—are not only unsound but also impoverish the experience and potential of earthly life for everyone. The exploitation of African land and labor has been more than a process of wealth-making by Western colonialists and development agents. It has also been integral to occidentalist perspectives of tropical nature, including ideas about the nature of poor people and postcolonial states upon whom so much blame is heaped for the vanishing of biodiversity.

For someone who has been involved in the practical side of conservation and development during my two years of Peace Corps service in the Comoro Islands (1989–1991), in California with the San Francisco Conservation Corps (1992–1993), and in Madagascar (1994–1995), when I carried out masters thesis fieldwork that had an applied dimension, I take a risk in presenting a work that evaluates conservation interventions critically yet avoids practical recommendations. I hope this will not be taken as an indictment of the pursuit of conservation but instead as a reflection of my sense of the futility of practical recommendations in light of current political-economic realities. Yet I am hopeful about the prospect of revolutionary solutions when the time is ripe.

My interest in Madagascar began during my three years (1993–1996) at Clark University, in Worcester, Massachusetts, where I pursued the masters in international development and social change, having every intention to return to Africa to continue working in development and environmental protection. Dick Ford suggested I apply for an IIE Fulbright grant for Madagascar, where he was involved as a consultant for an Integrated Conservation and Development Project (ICDP). Thanks to his contacts and guidance, and to the generosity of IIE Fulbright, I was able to return to the Indian Ocean to assist in developing a grassroots, participatory method of monitoring and evaluating the ICDP of the Andasibe-Mantadia Protected Area. This tool was meant to enable villagers to have not only stakes in but also some control over project interventions, such as tracking their progress in ways that were meaningful and legible to them. I worked on this, while also organizing with Hajamanana ("Haja") Rakotoniasy, my research assistant who ended up doing most of the work, a women's cooperative for selling woven raffia crafts. Yet I was also interested in collecting oral histories and landscape narratives from villagers, inspired by a riveting seminar on social forestry with Dianne Rocheleau, my thesis advisor. A talk by Arturo Escobar at Clark University spurred my interest in the greening of capitalism with respect to rain forest conservation. I thank them both for their sparks of imagination.

In Madagascar, I depended heavily on Malagasy friends and colleagues. I am especially indebted to Ndranto Razakamanarina and to my intrepid co-ethnographer, Haja, without whose friendship and support I would have been miserable. Haja and her family—her mother, Lalatiana, brothers Andry and Anjara, and sister Hoby—based in Moramanga, were truly my family away from home. As Haja and I would make the daylong trek back to the village of Volove after replenishing our provisions in town, we were routinely accompanied by two manual workers of the ICDP. Usually it was Theodore ("Bekapoaka") and Simon who shouldered much of our heavy load. Our conversations with them and other conservation agents gave me my first insights into the significance of low-wage labor in Madagascar's conservation effort, as well as into the implications of labor–management conflict. The experiences of the ICDP workers, particularly the event of a strike organized by the crew later, in 1996, sowed the seeds of this book.

On returning to Clark in 1994, Barbara Thomas-Slayter offered me invaluable opportunities in research, editing, and copublishing case studies on gender and development, and political ecology, for which I am in her debt. As I was writing my thesis, courses with Dick Peet on development and social theory and with Bob Vitalis on U.S. expansionism were probably most to blame for my turn away from the applied world of international development and toward academe. In 1996, I left for Baltimore to work with Gillian Feeley-Harnik in the doctoral program in anthropology at Johns Hopkins University.

Getting to know Gillian Feeley-Harnik made me realize how truly ignorant I was. Gillian's enthusiasm for ethnographic and historical detail and her ability to immerse herself in distant social-historical worlds must partly explain the origins of her special creativity. Conversations with Gillian intensify the wonder of everything. Her writings on political ecology and the religious iconography of landscape, human–animal entanglements, and the ethnology of work and labor in Madagascar have had a lasting effect on how I look at things. It was always a privilege to read her comments on my drafts, jotted on a ruffle of Post-It notes framing each page. Gillian gives so much time and inspiration to her students. I deeply appreciate all she has done for me.

Sara Berry's seminars in African history and historiography were crucial to the formation of my research project, and her mentorship was also invaluable. Funding from the Institute for Global Studies in Culture, Power, and History enabled me to return to Madagascar in the summer of 1997. Friends at Johns Hopkins who long after remained integral to my intellectual and social life include Kate Jellema, Roger Magazine, George Baca, and Elizabeth Dunn.

In 1998, I followed Gillian to the University of Michigan, where I found a community of Madagascar scholars, including Henry Wright, Conrad Kottak, and Gabrielle Hecht, and fellow graduate students Zoe Crossland and Nabiha Das. Needless to say, conversations with these people at different points benefited me greatly. Frederick Cooper and Jane Burbank's course on the comparative study of empire opened my eyes to the broader context of late-nineteenth-century European imperialism. Fred's book on "the labor question" in French and British Africa was seminal to my project, and I depended greatly on his incisive comments on my work. As concerns such ideas as the politics of memory and comparison, the aporias of colonial discourse and domesticity, and the anxieties of colonial governance that knot the grain of the archive, Ann Stoler's seminars created the conditions of the "aha moment."

My return to Madagascar from October 2000 until February 2002 was generously funded by the Department of Anthropology, Fulbright-Hays, and the IDRF Program of the Social Science Research Council, as well as the Program in Labor and Global Change of the International Labor and Industrial Relations Department at the University of Michigan. My gratitude to friends and colleagues in Madagascar is immeasurable. I particularly want to thank Jean Aimé Rakotoariso, director of the Institut des Civilisations / Musée d'Art et d'Archéologie of the University of Antananarivo, and Fulgence Fanony, head of University of Toamasina's Centre d'Études et de Recherches Ethnologiques et Linguistiques. For their warmth, generosity, and companionship, I owe a lifetime of thanks to Samoela Ranaivoson, Tiana Ranaivoson, and the inimitable Misa, who lit up my days. I relied greatly on the assistance and friendship of Zosy Gabrielle, my research assistant, as well as the translation skills of Billy and King, whose guidance was a godsend during those first months. Madame Monique Nireigna was also a welcoming presence and crucial resource for me. I owe so much to Very-Paul for his openness and assistance, and I can only hope one day he will be adequately rewarded for all he is and does.

Among a multitude of informants, I want to recognize those who spent hours talking to me and welcoming me to their hearths. They include Rakotoarisoa ("Tsimifira") Augustin and Mariette, Michel Akim, Mbosaña and Navony, Ernest Raveloson and Antoinette, Jaovita Paul, Soanette, Dely, Balbine, Samy, Pascaline, Paul Mahavantana, Tilahibe, and members of the biosphere personnel: Narcisse, Jery-Augustin, Georges Marcellin Randriamahefa, Paul-Réné, Feno Louis Dieu-Donné, Eugène ("Tovo") Ratovonera, Gilbert, Salim Ben Charif, Samby, Christophe Jean Josoa, Roland Emilien, Fidele Faustin, Jacqueline Mahaleo, Robertine Jeanne Rasoa, Jules Railison, and René de Roland Lylya. I thank Luc and Ries Toubert for company and generous use of their beachfront house when vacant. Thanks to Jennifer Cole for the bamboo bed and shelves that, despite a steady attack of powderpost beetles, gave sturdy comfort during my stays in Mananara-ville. I was overjoyed when two Peace

Corps volunteers, Robert Gronemann and Jennifer Tucker, were stationed in the vicinity. Having the occasion to spend time with compatriots over brochettes and Three Horse Beer was a welcome respite.

I returned to Ann Arbor in February 2002 to write. Individuals not yet mentioned who, then and later, offered comments, suggestions, or support at various stages of this project, or were otherwise indirect interlocutors, include Arun Agrawal, Nicole Berry, Sharad Chari, David Cohen, John Collins, Jill Constantino, Grace Davie, Paul Eiss, Juliet Erazo, Elizabeth Ferry, Crystal Fortwangler, Maria Gonzalez, Rebecca Hardin, Michael Hathaway, Karen Hébert, Larry Hirschfeld, Conrad Kottak, Erica Lehrer, Louise Lennihan, Ken MacLean, Erik Mueggler, Ed Murphy, Paul Nadasdy, David Pedersen, Maria Perez, Stephen Pierce, Ann Rall, Josh Reno, Doug Rogers, Audra Wolfe, and others whom I have unintentionally omitted and beg to forgive me. A shared subscription to organic produce and the beginning of a collective descent into baby chaos gave rise in 2002 to the Burdock group, including Jim Herron, Rachel Meyer, Jason Antrosio, Sallie Han, Jill Constantino, Michael Baran, Hank Wolfe, and me, by which a tradition of rotational hosting and communal feast-making was established. May it live on.

The reading groups and parties on Fountain Street, hosted by Fernando Coronil and Julie Skurski and animated by salsa music, were, from my perspective in those days, the life-pulse of our intellectual community in Ann Arbor. Fernando catalyzed politics into poetry and was committed to projecting a fairer world. He remained an important presence in my life up until the midsummer of 2011, when he was abruptly taken to an ICU in Manhattan and died several weeks later of metastatic lung cancer. One effect of a death is that it triggers the impulse to comb history for meaning, to ferret out clues of the shape of things to come, or to attempt to make light of darkness as "the owl of Minerva begins its flight." My mind is preoccupied by the devil in the details as I compare aspects of Fernando's life and death to the life and death by cancer of my mother, Inez Kemptner, in 2008. And I know that these cancers, and the extinctions that inspirit this book, came to pass too soon, on the brink of a moment in which they will likely be subdued or reversed.

Brighter has been the efflorescence of new life since my return from Madagascar. I met my future spouse, Hank Wolfe, in the summer of

2002 shortly after returning from Madagascar. Five years later, we had three beautiful children, Oscar, Vincent, and Scarlett. I will always be grateful for Hank's moral support and labors as this book took shape between and amidst the daily challenges of our careers and lively brood. By happenstance, my job brought us to Hank's home state of New Jersey, where his parents live still. I am indebted to Jacqueline and Neil Wolfe for their loving child care, and especially to Jackie, who even came to my office biweekly to tend to the successive newborns while I lectured. The former chair of our Sociology and Anthropology Department at Rutgers-Newark, Clay Hartjen, worked his usual magic and magnanimously lessened my teaching load during the semester my second son was born. In 2006, I was ecstatic to receive a Hunt Postdoctoral Fellowship of the Wenner-Gren Foundation, which gifted me the time to begin converting the dissertation into a book.

Under trying circumstances, my sister, Sheri, took in our mom as she grew frailer, which was such a comfort to me, as is having her there to share memories. And I thank my lucky stars for my father, Gary, and stepmom, Em, whose encouragement of my distant explorations, both geographical and mental, and whose love and support mean the world to me. During this last year of finalizing the manuscript, Johanna Rydström was a super-nanny and eldest "daughter," and we miss her awfully now that she's home in Sweden.

Sections of this book have appeared elsewhere in different forms. Two sections of chapter 4 are revisions of pieces of articles, including "Forced and Forest Labor in Colonial Madagascar, 1926–1936," *Ethnohistory* 52, no. 2 (2005): 407–435, and "An Exceptional Strike: A Micro-History of 'People versus Park' in Madagascar," *Journal of Political Ecology* 14 (2007): 10–33. Chapter 3 is an updated and expanded version of "Land and Languor: Ethical Imaginations of Work and Forest in Northeast Madagascar," *History and Anthropology* 15, no. 4 (2004): 367–398. A different version of a piece of chapter 5 appears in a volume I edited, *The Anthropology of Extinction: Essays on Culture and Species Death* (Bloomington: Indiana University Press, 2012). I thank these journals and IUP for permission to publish excerpts here.

Finally I want to acknowledge that the evolutions of the manuscript out of a form I surely would have cringed at to see in print were guided by the critical input of numerous readers. Although the remaining flaws are my own doing, I am very grateful to David Graeber and Christian Kull for their insightful and constructive suggestions. I thank Carol Kennedy and Marvin Keenan for their copyediting labors. And I owe a special thanks to my editor, Dee Mortensen, for believing in this project and showing such kind patience.

A WORD ON THE ORTHOGRAPHY AND PRONUNCIATION

The Malagasy language has dialectical variation throughout the island. In the book, Malagasy words reflect either the standard Merina spelling ("official Malagasy") or the Betsimisaraka pronunciation, depending on the context.

In the Merina dialect, the letter *o* is pronounced like the English *oo* (as in "zoom"). In the northern Betsimisaraka dialect, *o* is frequently pronounced long (as in the English "oh"). When I cite a Betsimisaraka speaker, I use the diacritical *ô*. Thus the Merina word *vola* (money) becomes *vôla* in Betsimisaraka, for example.

In Merina, the letter *n* is pronounced as in English, but the Betsimisaraka dialect frequently employs the velar nasal (the *ng* sound in "sing") where Merina does not. The velar nasal is conventionally written as *ñ* in the ethnographic literature of Madagascar. Thus *razana* (ancestor) in Merina becomes *razaña* in Betsimisaraka.

When citing the work of other Madagascar scholars, I copy the spelling they use for Malagasy vocabulary.

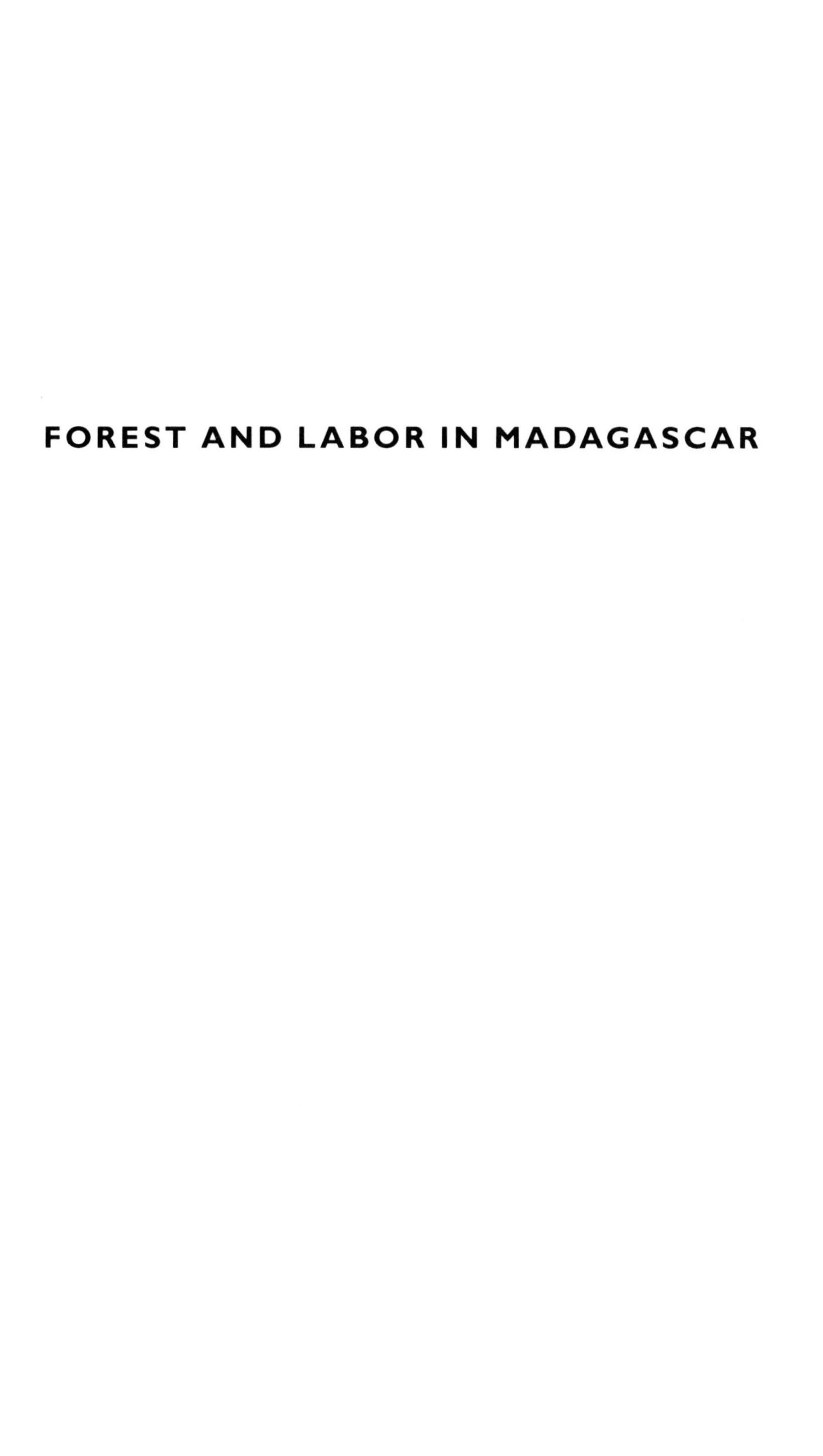

FOREST AND LABOR IN MADAGASCAR

It is therefore to this question of labor
that all other things are linked.

—Chef du district de Mananara, 1911

1 Geographies of Borrowed Time

On June 13, 2010, a story about the plunder of rosewood trees out of several national parks in Madagascar made the cover of the *New York Times* (Bearak 2010). Not only Malagasy citizens but also international readers concerned about biodiversity protection had been following the story for several months, ever since the coup d'état of the previous March that ousted the pro-conservation and pro–United States president, Marc Ravalomanana, and left Madagascar's hinterlands open to a new scramble for Madagascar's untapped resources: A new wave of imperialist expansion, now launched from the East rather than Europe.[1]

A ring of Chinese and Malagasy merchants, dubbed the "rosewood mafia" in news reports, armed gangs of "thugs" to intimidate residents and park guards around the rain forests that line the northeastern Antongil Bay. The "Timber Barons," as they are also called, having Malagasy names such as Bematana, Bezokiny, and Body, and Chinese ones such as Chan Hoy Lane and Sam Som Miock, infiltrated major towns on the east coast (Wilmé et al. 2009). They hired local villagers for dirt-cheap wages and shipped in extra hands from "deep China" (Gerety 2009a). North American and European expatriates were flown out to safer havens. Conservation activities ceased while local officials, colluding with the Timber Barons, gave the loggers free rein in the national parks.

These activities continue as I write this introduction in spring of 2011. The work gangs of rosewood loggers forge deeper into the forests of Antsiranana Province. It all started at Makira, a national park that was created in 2006, marking a triumphant moment for conservation advocates in Madagascar who had for years been pushing for the expansion of protected areas. Back then when he was a few years into his term, President Ravalomanana promised to do just that. But with devastating irony, the timber merchants, not the conservation organizations, seized the remnant isles of rain forest to convert timber, rather than the experience of tree-filled parks, into cash.

Gangs of loggers fell the majestic trees with handheld saws, then roll them over the steep and knobby forest floor. They lash them to rafts and float them downriver toward the ports.[2] Most of the timber is shipped to China, feeding the desire of a growing Chinese middle class for Ming dynasty reproductions (Garety 2009).[3] It is also coveted for its sonic properties, for the "thickness and creaminess" it lends to the tone of a Gibson guitar (Hunter 2007).[4] Some enterprising types have sought extra money poaching lemurs, birds, tenrecs, and other game from the national parks. In 2010, photographs of a pile of taut and blackened lemur corpses were posted on environmental websites (Bourton 2009). Although it was reported that restaurants in Madagascar sell this new delicacy, I later learned that most of the meat is consumed by loggers sleeping in rough encampments in the park's interior. For Western readers, the images of endangered species–turned–bush meat add ghastly detail to the rosewood debacle, bringing to mind Edward Said's insights that cultural difference, repulsive or strange to the Westerner, need say no more to establish otherness. Through implicit juxtaposition, the Orient can be "made to serve as an illustration of a particular form of eccentricity . . . a grotesquerie of a special kind" (Said 1979:103).

Fairly rapidly, the loggers moved southward to Mananara-Nord, my field site, where over the course of fourteen months between 2000 and 2002, I traced the steps and recorded the words of resident Betsimisaraka men employed by an integrated conservation and development project (ICDP) at one of UNESCO's global biosphere reserves. Current events have put my historical ethnography of conservation labor in a different light and grammatical tense than it might have otherwise been were it not for the incursions of the Timber Barons and the interruption of conservation activities there. Insofar as a historical ethnography of forest conservation and low-wage labor helps to make sense of a particular situation in Madagascar, it may also help make sense of why editors at the *New York Times* might assume that their readership would

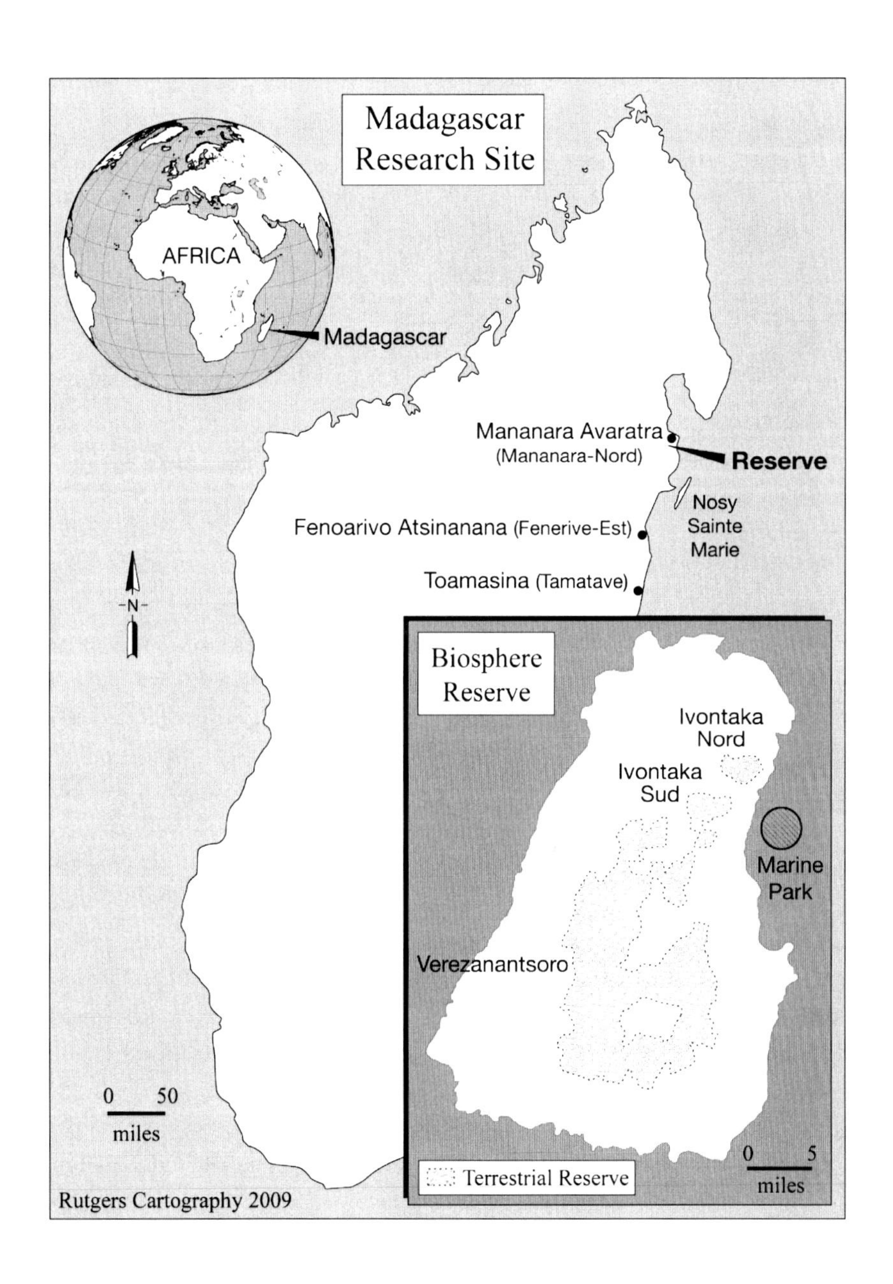

Map 1. Map of the Mananara-Nord Biosphere Reserve, Madagascar.
Created by Rutgers Cartography Services, 2009.

care about the looting of Malagasy rosewood and poaching of endangered lemurs.

This book charts a materialist history of biodiversity conservation around the Antongil Bay from the beginning of the twentieth century to the turn of the twenty-first—that is, from the onset of French colonial rule in 1896 to the beginning of Ravalomanana's ill-fated presidency in 2002. It grounds the discourse of species loss and salvation in the tasks and structure of conservation work. At the center of this story is a class of laborer known as a "conservation agent," the official title given to manual workers of ICDPs in Madagascar during the time of conservation's resurgence in the mid-1980s.

Conservation agents are responsible for the physical tasks of protecting biodiversity. Over the course of my stay in the Mananara-Nord Biosphere Reserve, conservation agents of Mananara-Nord, all of whom were male, monitored the boundaries of the biosphere reserve's protected areas, including a marine reserve and a rain forest reserve. They reported on rule-breakers, catalogued species inside the park, groomed footpaths, built park infrastructure, disseminated conservation rules and practices among villagers, and eventually became guides of tourists and foreign scientists in the reserves. I compare conservation agents to their structural counterparts of French colonial rule (1896–1960), an era that had been primed by the internal colonization and subjugation of most of the island by Merina rulers until their defeat by French forces. The Merina, as well as Betsileo people, are Malagasy ethnic groups whose natal territory comprises the High Plateau region. Since Madagascar's independence from France in 1960, Merina people have virtually monopolized state offices, even though former president Didier Ratsiraka, who held on to state power for nearly twenty-five years, is Betsimisaraka.

The scramble for precious hardwood along the east coast has made it unsafe for the conservation agents I knew to do their work. I imagine them now tending to their own farm plots, gardens, and households, waiting things out while the coup regime headed by Andry Rajoelina remains unrecognized by other nation-states and complicit in shaping the volatile atmosphere of illegal logging. Conservation representatives of Madagascar are optimistic that order will be eventually restored, but the ecological damage inflicted so far will have further jeopardized innumerable species' lives and human livelihoods. What will become of

Madagascar's diverse habitats and societies over the next decade and beyond will depend on the nature of the new state, on the state of ecological devastation, and on the interest, trust, and labor time of rural people who live on what Anna Lowenhaupt Tsing (2005:32) calls the "salvage frontier," a space "where making, saving, and destroying resources are utterly mixed up, where zones of conservation, production, and resource sacrifice overlap almost fully, and canonical time frames of nature's study, use, and preservation are reversed, conflated, and confused."

By examining in one place the exchange between people and nonhuman nature, this book throws light on a general process that takes place everywhere, but always under particular conditions and in specific modalities. This is not just a history of the making of nature through labor or of labor through nature, but one that enjoins the mutual formation of people and nature through the process of transforming a specific space at a particular time. This takes not just time, but time formed by a specific history of a particular place.

LABOR AND PROTECTED AREAS

In the mid-1980s, Western donors launched a plan to protect the island's biodiversity through economic liberalization after nearly a decade of socialist policy. Donors' prioritization of biodiversity conservation and sustainable development transformed Madagascar's job market. Well-educated job seekers with skills in agronomy, ecology, participatory methods, and community development found employment in a growing array of international NGOs and agencies devoted to conservation and development.

The international interest in Madagascar's biodiversity loss was not new. Forest and soil conservation had preoccupied the French colonial state since at least the 1920s, and even earlier if one takes into account Governor-General Gallieni's experimentation with the cultivation of native tree species soon after he arrived in Madagascar in 1896 to organize the new colony. French officials were already seeing the ecological devastations of projects of *mise en valeur* ("valorization," or bringing land under capitalist production). Conceding the follies of repressive conservation policies that mimicked colonial-era efforts and repeatedly failed to get peasants to stop slashing and burning the rain forest to grow rice—a practice known in Madagascar as *tavy*—neoliberal planners rejected a unilateral, top-down model of priority-setting in favor of a more people-friendly approach. This consisted mainly in meetings between conservation project workers and villagers, where villagers could voice their preferences in the services (e.g., veterinary, health, nutrition) they desired or the structures they wanted to see fixed or built, such as bridges and schools. In exchange for these local development activities, in which villagers were expected to volunteer their labor and provide a portion of building materials, villagers would agree to respect the rules forbidding them to clear forest or harvest timber or endemic species from legally protected areas.

By and large, Malagasy peasants have not, over the past twenty-odd years, applauded the new approach to conservation, especially after the funds from tourist ticket fees that were supposed to go toward community development in the early 1990s were slashed, once the state and donors realized that they could not afford to give up 50 percent of tourism revenue for the nascent park service. With some exceptions, notably through a hopeful initiative in community-based forest management, Malagasy peasants on the east coast have continued to resent and resist conservation efforts. This has made the employment of local residents as conservation agents problematic. Locally hired people, as insiders, possess knowledge and social connections that are essential to the conservation effort, yet their sense of duty to the ICDP is often compromised by their loyalties and pragmatic strategies to offset the economic insecurity of conservation work with subsistence labor.

In tropical regions such as Madagascar, scholars and conservation practitioners have identified and sought to assuage conflicts between local populations and environmental projects. The scholarship on "people and parks," or conservation interventions in the global South, flourished shortly after big development institutions repackaged foreign aid to conform to the vision of environmentally sustainable development outlined in *The Brundtland Commission's Report* (1987) (see Neumann 1998; Harper 2002; Brockington 2002; Igoe 2003; Walley 2004; Haenn 2005; Lowe 2006; West 2006; West et al. 2006). Anthropologists Paige West, Jim Igoe, and Dan Brockington (2006) synthesize and draw out the common themes of studies of the people-parks dynamic up to their article's publication date. These have been significantly informed by the theoretical perspective of political ecology, what Aletta Biersack succinctly defines as an approach investigating "how power relations mediate human-environment relations" (Biersack 2006:3). The by-now familiar framework of "people versus parks" has illuminated diverse sources of conflict between conservation authorities and "targets" of policy interventions, as well as the negative social effects of conservation schemes in biodiversity hot spots.

A burgeoning literature in social science turns attention to the relationship between capitalism and environmental protection, as featured in a 2010 issue of the geographical journal *Antipode*. Such studies examine the emergent partnerships between private corporations and conservation organizations, the political significance of capitalizing intact or semi-restored landscapes, and the application of a capitalist logic to the mission of biodiversity protection (MacDonald 2010). Newer studies also document the trend in ecological economics of setting monetary values on ecosystemic services thereby strategically submitting conservation to "'free market' processes" (Igoe et al. 2010:488). This includes analyses of the social and economic dimensions of the global initiative Reducing Emissions from Deforestation and Forest Erosion (REDD+), launched in 2005, which entails an international cooperation in mitigating the effects of carbon and other emissions attributed to the disappearance of forest.

My intervention into the political ecology of conservation centers on the role of manual labor in creating the value of endemic tropical species within industrialized metropolitan centers. Labor is a conceptual lens through which to examine the effects of hierarchy, differential compensation, resistance, and acquiescence on the creation of value. I concentrate on the people at the lowest levels of the social hierarchy in Madagascar—what I call subaltern labor—to expose the mundane tasks that have made possible the acquisition of certain types of knowledge, and the evolution of certain philosophies of nature. The workers who do all the grunt work of protecting animals, plants, coral reefs, and rain forest from extinction have been virtually invisible in accounts of what has failed and what has worked in conservation efforts. Their obscurity reinforces the view that the conservation of nature, like women's domestic labor, like Mother Nature herself, is "an antithesis of human productive activity" (Smith 1990:368). The idea that conservation is a palliative to extractive activities in the rain forest has only shrouded the contributions of subaltern workers who have prepared the grounds for Westerners' romanticization, exploitation, discovery, and salvation of tropical wilderness (see White 1996:171; Slater 2002). This idea of conservation furthermore assumes the preexistence of intrinsic value in nature: conservation does not create value but protects an a priori value and accumulates it by enabling biodiversity to proliferate.

In the early 1990s, the conservation agent, a locally hired extension agent of sorts, represented a new kind of worker-peasant in peripheral, tropical economies. This class of worker-peasant, like a civil servant, had to enforce environmental legislation, but more than that he was expected through his words and deeds to spread conservationist ideology to members of his own ilk. The responsibilities were all out of proportion to the actual numbers of conservation agents employed by ICDPs. At Mananara-Nord over the course of my research, there was a median of ten men (accounting for resignations and the time lag of new hires) to patrol 140,000 hectares of biosphere reserve. The reserve was overseen at the time by a Dutch representative of UNESCO and his Malagasy counterpart, whose title was national director. The site was the first of UNESCO's biosphere reserves in Madagascar, as well as the first ICDP on the island.

Manual conservation workers, trying to make ends meet with wages that do not in themselves cover household food needs and expenses, while also pressured to maintain harmonious relations in the village, and always threatened with the sudden loss of employment, had to stay active in the subsistence economy of *tavy* as recipients of rice cultivated by kin members, or as cultivators themselves. The contradiction of transnational conservation efforts in Madagascar is that emplaced, low-wage workers engage in the moral economy of *tavy,* the land use most blamed for whittling away the rain forest.

The analysis of conservation as productive labor attends to the tensions internal to the labor structure of neoliberal conservation and development efforts, thereby shifting attention askance of the "people versus park" dynamic. I am interested in the ways in which rural Malagasy workers have pushed back against what they view as overreaching by "outside" capital, including conservation and development entities. Yet these agrarian workers that have been incorporated into neoliberal conservation and development have also guardedly assimilated key principles of conservation, since they are directly impacted by ecological changes, such as the increase of flooding, the depletion of soils, land scarcity, and the decrease in wild protein sources. They also recognize that many of today's conservation and development practitioners (the expatriate experts and project managers, and the Merina consultants and extension agents, for example) are aware of and sympathetic to the constraints faced by the rural poor who practice *tavy.* The focus on conservation labor therefore underscores the fact that Malagasy conservation agents are not merely sympathetic to rural populations but are also in the same boat.

Marx's labor theory of value orients the book—that is, the thesis that labor (socially necessary labor time) produces commodity value in the capitalist mode of production. I argue that the devaluation of manual conservation labor relative to intellectual labor, and relative to the endangered, endemic species themselves, enhances the global value of endangered biodiversity, even though it thwarts the goal of preventing eventual extinctions due to habitat loss. Devaluation here refers not only to meager compensation for arduous work relative to intellectual labor and in accordance with one's ethnic or national identity. But it is also tied to the redeployment of a historical moral hierarchy that maintains the imbalance between unskilled, "coastal" labor and elite others in Madagascar. The production of labor's value in the conservation and development bureaucracy is inversely related to the value of labor's product, which is, in large part, an accessible, scenic park more densely stocked with species life than it otherwise would have been without surveillance.

The inverse relationship of value between low-wage conservation labor and protected yet threatened wilderness suggests how tropical parks have been fetishized in Western societies. But the most glaring problem of the devaluation of manual labor is that it fixes workers' attachment to the very land use and land ethic targeted by conservation organizations for rapid, radical change: namely, "slash-and-burn" agriculture (*tavy*). Getting Malagasy peasants to stop practicing *tavy* in primary forest has been the cornerstone of the global effort to stabilize the island's biodiversity loss. The fact that the manual conservation workers have themselves practiced *tavy* and other forms of subsistence complicates their duty to spread the word and deeds of conservation.

MADAGASCAR'S TRANSFORMING NATURES

The tragedy of species loss in Madagascar for scientists and environmentalists lies in the island's ecological variation and uniqueness. Madagascar contains "relics of a vanished geological age," its plants and animals having evolved in isolation when tectonic plates shifted around 160 million years ago and severed the island from mainland Africa (Grandidier 1920:197). This accounts for the island's extremely high degree of species endemism. Between 80 and 90 percent of its plants and animals exist nowhere else on Earth.[5] Identified by scientists as a "biodiversity hot spot" in the late 1980s, Madagascar and its forest biomes possess "exceptional concentrations of species with high levels of endemism" and an unusually high rate of depletion (Myers 1988).[6]

Over the past two and a half decades, habitat degradation and biodiversity loss have forged global representations of Madagascar as "a bleeding island," a "biodiversity hot spot," and a zoography on the "brink of extinction." Rapid habitat loss and rising species deaths innervate us and speed up our perception of time. While metaphors of Madagascar are premised on specific biogeographies and temporal frames, their allusion to the unpleasant sensations of hemorrhaging, plummeting, and combusting intend to stir and upset us. They reflect values of particular ecologies as much as they project existential anxieties.

The erosion of the island's rain forests had already begun before the arrival of European settlers, but deforestation increased dramatically after French colonization.[7] In the past half-century, roughly half of the forest cover has vanished (Hanski et al. 2007). Satellite data of Madagascar from the late 1980s and early 1990s show that 66 percent of the original primary forest of the eastern humid zone had been burnt, mostly for the purpose of rice cultivation. Mining, logging, and road and rail works have also contributed to the decline of the forest.[8]

From the vantage point of conservation advocates, the current rosewood plunder in Madagascar has temporarily usurped *tavy* as the main scourge of the eastern forests (Jarosz 1993; Kull 2004). In 1896, when France annexed Madagascar to its empire, the colonial state banned *tavy* and commercialized rice, the staple crop, in order to initiate a process of valorization for both Malagasy nature and labor (Feeley-Harnik 1984; Jarosz 1993). Before France's conquest, the Merina monarchy claimed ownership of the island's primary forests but rarely intervened when local areas exploited their natural resources. Traditional taboos (*fady*) dictated which forest areas were off-limits to burning or razing (Ramanantsoavina 1966). A year after formal colonization, in 1897, the French state established the first regulations concerning forest conservation for the purpose of capitalist exploitation, including mining, plantation agriculture, and logging by entrepreneurs. These concessionaires, as they were called, were granted forest parcels (concessions) from the state (You 1931: 406). In a relatively short time, the problem of deforestation preoccupied colonial foresters, since infrastructure development, industrial production, and *tavy* were decimating the island's biodiversity, a fact much rued by scientists and amateur natural historians at the time.

Geographer Christian Kull (2004:180) argues that the state's criminalization of land burning and peasants' resistance to it have produced "complex entanglements of power" that have made it impossible to conceptually segregate officials from peasants, or criminal acts from acts of resistance. The state and the peasantry are mutually constitutive, intermeshed parts of a broader society, itself shaped by the complex power arrangements among the nation-state, Western governments and organizations, and private capital. As such, the class, caste, ethnic, political, and ethical interests of foresters, state officials, peasants, NGO workers, donors, ICDP directors, and so on overlap and conflict, depending on context.

Colonial projects of valorization in Madagascar, and throughout Europe's colonies in the nineteenth century, started with decrees that cordoned off land for industrial extraction, state concessions, and, eventually, nature conservation. Land enclosure, taxation, and forced labor comprised a three-pronged strategy to develop African labor forces for capitalist production. As peasants and pastoralists were alienated from their means of subsistence, they were also severed from the natural and symbolic resources that formed the bedrock of group identity and history (Neumann 1998).

In Madagascar, the French state legislated public works to build transportation and communication networks, and it welcomed entrepreneurs in the logging, mining, and plantation industries. Both the state and private capitalists despaired of an insufficient labor force in Madagascar. Malagasy peasants had no desire to work for their colonizers, and preferred to escape deep into the forested mountains, when possible, to escape compulsory state labor or the collection of head taxes. For individuals who could not evade *corvée* labor or taxation and found themselves laboring on roads and railways, or on private landholdings when they were "borrowed" by concessionaires in search of manpower, project overseers expected them to provide their own meals (Sodikoff 2005a). Malagasy workers relied on their spouses or extended kin networks for food, and kin members generally farmed rice in proximity to work sites. In areas where road and railway construction was penetrating unsettled rain forest, people practiced *tavy* to plant rice and other food crops. Although officially banned, the necessary persistence of *tavy* benefited the growth of capitalist production because it cheapened labor.

In addition to the pragmatic interest in Africa's natural resources, Europeans held romantic preconceptions of Africa as a vast territory of relatively unsettled rain forests, savannahs, mountains, and wetlands: sites of exploration, adventure, contemplation, and leisure. Nature reserves created to protect animals, soils, and unique habitats became sites in which Europeans could hunt game and savor the primordial scenery (Anderson and Grove 1987; MacKenzie 1988, 1990; Neumann 1997; Brockington 2002). The colonial transformation of African geography emulated what happened in the United States, where national parks came into being through the violent displacement of Native American populations (Croll and Parkin 1992; Escobar 1999; Neumann 1998; Jacoby 2001).

In the 1920s, French botanists Henri Perrier de la Bâthie (1921) and Henri Humbert (1927) theorized that a continuous "evergreen forest paradise" once blanketed Madagascar, and that this forest cover gradually had shrunk since the arrival of the first human settlers from Austronesia. This occurred approximately 2,000 to 2,500 years ago, according to evidence based on the charcoal stratigraphy of sediment cores (see Burney 1997). This colonial narrative not only erroneously depicted the precontact landscape of Madagascar, which was more likely a patchwork of grassland and forest, but it also absolved colonial settlers of environmental wrongdoing by their extractive enterprises in the Malagasy forests (Pollini 2010).[9]

The state stepped up conservation activities in the mid-1920s while continuing to support industrial production in the eastern forests (Kull

2004). By 1927, the colonial state delimited 353,597 hectares of forest as ten nature reserves, placing them under the control of the Natural History Museum of Paris. The move signaled a growing anxiety about the colony's rapidly diminishing forest domains.[10] Meanwhile, policies requiring Malagasy subjects to find wage work put multiple pressures on agrarian populations. It is no wonder that they resented the hypocritical and repressive conservation efforts of the state.

Despite the colonial state's deepening commitment to soil and forest conservation over time, particularly after World War II, its conservation policy was differentially implemented. The ban on *tavy* continued, yet so did the extraction and commoditization of forest resources (Olson 1984; Jarosz 1993). Pierre Boiteau (1958:225) writes that in 1954, six years before Madagascar's independence, the French state had expropriated around 10 million hectares of primary forest in Madagascar, about 9 percent of which it classified as integral reserves (*réserves intégrales*), strictly off-limits for any purpose other than authorized scientific research. Concessions constituted the vast percentage of forest domains, which the state exploited or leased to plantation owners, loggers, and miners. Eventually, the state's interest in soil conservation and *mise en valeur* was overshadowed by the problem of Malagasy resistance and the growing nationalist and labor movements that emerged in response to the draconian labor laws of the French state.

The French finally conceded control over Madagascar to the nationalists in 1960, when Philibert Tsiranana took power. Like other leaders of newly independent African nations, Tsiranana maintained ties to France as well as colonial legislation. Amidst growing discontent with his administration and disillusionment with independence, Tsiranana resigned the presidency in 1972. This marked the beginning of a socialist revolution that culminated in 1975 with the rise to power of Lieutenant Commander Didier Ratsiraka. The socialist era redirected the state's attention to the pursuit of agricultural self-sufficiency. Forest conservation, associated with colonial repression, was not a priority.

But looking back in time again, to the period when banning *tavy* was a central preoccupation of the colonial state, I reflect on one reason why the colonial state's harsh tactics in suppressing *tavy* in the forest failed. Resistance to colonial authority was a vitally important factor, but so was the state's tacit encouragement of *tavy* as a means of minimizing costs for public works and private industries. I scrutinize this inconsistency in colonial policy because it has persisted within neoliberal conservation efforts, which one could see as a kind of public works campaign in that conservation and ecotourism development are presented as national and global public goods. They also entail the solicitation by ICDPs of agrarian people's voluntary labor.

Up until my departure from Madagascar in February 2002, ICDPs were hiring residents of the surrounding protected area of a national park, the Mananara-Nord Biosphere Reserve, who even after their employment continued to participate in cash cropping, petty trade, and subsistence agriculture to supplement their ICDP wages. The division of conservation labor, the unfair apportionment of due credit and reward, the rigidity of the hierarchical structure alongside the attempt to impose order on the unruly "workplace"—the national park and surrounding settlements—were a counterforce against conservation agents' desire to acquire species knowledge, learn foreign languages, teach others to adopt conservation practices, and see the region develop.

GREEN EXPECTATIONS

Scientists and policy makers have been concerned about the problem of global biodiversity loss since 1970 (Orlove and Brush 1996:329). But not until the mid-1980s did financial aid institutions begin to invest heavily in biodiversity conservation in underdeveloped tropical countries. This was the dawn of what Michael Goldman (2004:166–167) calls "green neoliberalism" in the global South, when the World Bank and its finance partners pronounced free-market solutions to environmental problems and established certain truths about who was to blame for degradation and who would be in charge of righting wrongs. Globally, forest areas were diminishing by about 13 million hectares per year, and thousands of species were vanishing (Kaufmann 2008:34).

The surge in environmental funding in Madagascar changed the role of environmental nongovernmental organizations (NGOs) from advocacy and activism to managerial entities of ICDPs within a sprawling environment and development bureaucracy (Gezon 2000:184). The influx of foreign aid engendered a proliferation of international and domestic NGOs to head ICDPs in different protected areas, a trend that has been deeply explored by anthropologists of Madagascar (Gezon 1997a; Harper 2002; Kull 2004; Hanson 2007; Kaufmann 2008; Keller 2008). Jobs with these NGOs are considered by Malagasy workers to be very desirable because, although they exist for a limited duration, they offer better work conditions and compensation than many of the factory jobs in the garment, fishing, pine export, and mining companies of the free trade zone, as well as the more recent illegal logging operations of the Chinese and Malagasy export firms.

With the rising influence and role of NGOs during the neoliberal turn in postcolonial (and often postsocialist) African nations, ecotourism development and biological prospecting surged in the tropics. Estimates for the global number of ecotourists (broadly defined) range from "157 to 236 million, generating expenditures up to US$1.2 trillion" (Ceballos-Lascurain 1996:46–48 cited in West and Carrier 2004:483). In Madagascar, tourists must pay park entry fees, which are the main source of revenue for the national park system. For a foreign tourist, the tariff to enter one of Madagascar's national parks is US$12.50 per day (capped at US$25 for a four-day visit). Revenue from ecological and "pro-poor tourism," which includes stays in rural villages where tourists can get a taste of authentic Third World life, was on the rise, at least shortly before the March 2009 coup d'état (Duffy 2008:332–333).[11]

For international visitors, tropical nature reserves offer escape from the work grind of industrial society and national natures seen as "under threat from the profit-seeking growth that drives corporations" (West and Carrier 2004:483). The desire of the North American, European, or Japanese workers to get away from it all in the underdeveloped and verdant margins of the global economy expresses the irony of environmental nostalgia, a longing for a preindustrialized nature that modern industrial life instills (Frow 1997). Ecotourism offers a means of achieving growth without environmental destruction. Intact rain forest, long a lure for European adventurers, natural historians, and fortune hunters, was commoditized in such a way as to avert destructive transformation, or the process by which raw "nature" gets converted into "culture" (Ferry and Limbert 2008).

Rain forest parks do not need to be excessively fiddled with to have touristic appeal. They mainly need to be made navigable, their assets highlighted through well-laid footpaths. The nature reserve is a kind of commodity that breaches the conventional manufactory stage of commoditization. Ecotourism thus offers an instance of the "time-space compression" that defines the novelty of late-twentieth-century capitalism compared to earlier forms, according to David Harvey (1990). Time is felt to be "running out" as the rate of habitat and species loss speeds up. In the remote space of the rain forest park, the green capitalist mode of production accumulates surplus value less by the metric and labor-discipline of clock time, which is nearly impossible for authorities to regulate in hard-to-access rain forest parks, than by the ticking of an invisible "doomsday clock," the end result of uncontrolled habitat erosion and species loss.

The spirit of the Wild West that possesses industrial loggers and poachers on the east coast today has helped to deepen Madagascar's "extinction debt," or the time lag between an environmental perturbation and the species deaths that inevitably result, though the exact times cannot be determined (Tilman et al. 1994). The strained relationship between "ecological capitalism" and "extinction debt" hinges on this problem of time lag: on the one hand, planners envision the fruits of conservation and tourism eventually providing sustenance without degradation; on the other hand, the future death of species marked for extinction might appear to negate the results of present-day biodiversity protection, further embittering Malagasy people toward conservation representatives, who urge them to sacrifice today for the good of tomorrow.

The greening of development in the global South is built on the hope that nonextractive forms of wealth creation will liberate heavily indebted countries by putting them on the path to sustainable economic growth. In the early 1990s, even conventionally extractive industries, such as mining, began to go green. An example is the case of Rio Tinto, the British and Australian mining conglomerate. With the help of a $35 million loan from the World Bank, Rio Tinto extracts ilmenite in southern Madagascar. Ilmenite is smelted and refined into titanium dioxide, a base pigment in paint, paper, and plastics, most of which is sold to China (Revill 2005). To show they care about the resulting biodiversity loss that comes with mining in primary forests, Rio Tinto representatives have drawn up plans to rehabilitate defunct mines by using "a floating dredge to pull up deposits of sand." The dredged-up deposits will provide the foundation for ecological restoration; the deposits will be reseeded "with key species the company found in the original forest" (Gerety 2009b). Rio Tinto promotes its "Net Positive Impact" on biodiversity, boasts of its "Conservation Areas," "Conservation Committee," "Ecological Research Centre," and trials of "ecological ecosystem restoration." Their commitment to expressing environmental interest belies the fact that mining operations do not create a "net positive impact" on biodiversity but instead ravage forest. Sian Sullivan (2010) argues that environmental crisis and the search for its solutions have themselves become sources of growth for capitalism, engendering products and services such as "offsetting, payments for ecosystem services, natural capital, green-indexing, biodiversity derivatives, green bonds, [and] environmental mortgages" (Sullivan 2010:5).

These developments support the position of scholars, such as Martin O'Connor (1994a) and Arturo Escobar (1996), who hold that the ecological phase of capitalism is primarily a change in capital's representation of itself. Competitive advantage "is solved not by changing the means of production but by changing how meaning is produced, or how the relationship between persons and things is construed and managed" (Foster 2008:10). Nongovernmental and nonprofit entities play a significant role in establishing the relevance of green capitalism through the circulation of texts and imagery about habitats, species, and maladaptive land-use practices, as well as the denomination of objects of value by donor agencies and multilateral institutions. These organizations employ people themselves, and also have influenced the ways in which conventional capitalist industries operate and project their image to the public. It is about averting a crisis at multiple points. Ecologically oriented industry or environmental remediation activities by industries reflect capital's attempt to avert a crisis that originates in production conditions. Ecological Marxist scholars call this kind of predicament the "second contradiction of capitalism" (see contributors to M. O'Connor 1994b; J. O'Connor 1998; Benton 1996), which differs from the first contradiction, that of the demand-side crisis arising out of the tensions between capital and labor. As capital seeks to increase the amount of surplus value it can extract from labor, larger numbers of workers are either laid off or paid less relative to the task load. Consumers therefore buy less, and capital's profits suffer as a result. The second contradiction stems from the supply side, James O'Connor explains. This can happen when the material conditions of capital's production are poorly maintained, for instance, through the deterioration of workers' health, soil fertility, or infrastructures such as roads, ports, and machinery. It can also arise when public outcry and social movements impose demands that raise costs for capital or reduce its flexibility (J. O'Connor 1994:162). Attempts to resolve problems of the second contradiction have included the restructuring of the land-labor relationship, the transformation of industries' self-representation (e.g., "greenwashing"), and the delivery of better benefits or compensations.

Martin O'Connor argues that "second contradiction" crises are typically diagnosed as management problems, and these problems in turn can present another source of dynamism for capitalism (M. O'Connor 1994a:128). Capital's proclivity to make hay out of environmental crisis is exemplified by the development of ecotourism and rent-generating scientific study in nature reserves, all of which is included in today's conservation effort. While planners have expected these ventures to improve the lots of the rural poor over the long term, they often resuscitate colonial "structures of dominance" (Stoler 2008:193) as well as the "structures of feeling" (Williams 1977) that accompany colonial exploitations. The latter often manifest in rural tropical societies as a suspicion that outsiders' anxiety over biodiversity loss and the impoverished future is a ruse to dispossess the poor of what is left of their fertile, forested land.

Biodiversity, which the United Nations conceptualizes as the "natural heritage" of any given nation-state, constitutes the inner essence of a place (see Collins 2008). Nature's "intrinsic value" is thought to lie in its biological essence and outward form. It lacks artifice; it exists independent of human productive activity. Elizabeth Garland (2006, 2008:63), an anthropologist whose interests are akin to my own, has studied the role of African game wardens in conserving wildlife in Tanzania. She argues that conservation "relies heavily upon ideological mediation to add value to the initial natural capital that game animals represent. . . . [W]hat is crucial to the symbolic and economic capital produced by conserving African wildlife is the image of Africa . . . as a privileged space of nature within the global symbolic imaginary" (Garland 2008: 63). As a form of production, ecological conservation shrouds its own artifice, as workers suppress unsightly productive activities that historically both have lowered labor costs for capital and also, due to their cumulative land-degrading effects, have given rise to conceptions of "intrinsic value," "natural heritage," and "brink of extinction."

THE ETHNOGRAPHIC CONTEXT

The events that led to the rosewood looting began to unfold six years after I departed from Mananara-Nord in 2002, having spent the previous fourteen months in villages incorporated into a UNESCO biosphere reserve.[12] Mananara-Nord is the name of both a town and a larger prefecture on the lower lip of the Antongil Bay. The Mananara-Nord Biosphere Reserve encompasses 140,000 hectares of rain forest and marine territory as well as more than two hundred traditional hamlets comprised of Betsimisaraka *tavy* farmers and fishermen. Since the biosphere reserve's inception, peasant opposition to conservation efforts has occasionally erupted, but with nothing like the violence of the illicit logging and animal poaching described in recent news reports.

The prefecture of Mananara-Nord is hard to reach by the dilapidated Route Nationale 5. When I arrived in October 2000, it was less populated and inundated with cash crop buyers than the fertile districts that lie farther south, where it is easier to get crops to the road and to the seaport of Toamasina (Tamatave, in French). In the early 2000s, Mananara-Nord residents pursued a variety of livelihoods, including agriculture; service labor (in locales such as small restaurant-bars, hotels, and hair salons, or working as street vendors or *taxi-brousse* drivers); petty commerce in inexpensive imports, frippery, and furniture; crafts such as cabinetmaking and crystal artisanship; building; commodity production (such as crops, fish, meat, clothes, and crystal decorations); civil service; and domestic work, particularly for the relatively well-to-do Chinese, Merina, and Indo-Pakistani households in town. Several large companies in the import-export business for cash crops were based in the main town of Mananara-Nord, called "Mananara-ville" to distinguish it from the prefecture name. They traded primarily in cloves, vanilla, and coffee. One such company, the Malagasy-owned Ramanandraibe, possessed a large store in town that contained imported electronic goods and a ticket counter for Air Madagascar. Although Mananara-ville had a small airport, it was and is still perceived by city dwellers as a remote outback. The town and its peripheral villages had a population of about 35,000 according to a 2005 census.

The wealthiest residents of Mananara-ville included assimilated Chinese merchant families, a minority of Indo-Pakistani merchant families, and Merina families transplanted from the central highlands. The town was comprised mainly of wood-plank and corrugated-iron houses, but the wealthy merchants owned two-story concrete houses with glass windows and balconies. These dwellings, second homes really, were modest compared to the owners' palatial homes in cities such as Toamasina and Antananarivo. The Chinese dominated local commercial activity, including the trade in retail goods and the collection and shipment (by boat) of lychee fruit, vanilla, cloves, and coffee to the port city of Toamasina. The Indian families were also involved but had a less visible presence. Merina residents in town considered themselves a community in exile, and most were posted to Mananara-Nord as civil servants (teachers and doctors). I lived with a Merina family in town, the uncle, aunt, and cousin of my friend and former research assistant, Haja. They earned income from sewing and processing quartz crystal into polished balls and decorative objects. I stayed with them between my stints in villages surrounding the Mananara-Nord Biosphere Reserve.

Numerous wood-plank houses stood at the outer limits of the town along several arteries leading into the town center. On the outskirts of town began "the countryside" (*ambanivôlo*, "under the bamboo," pronounced *ambanivolo* by Merina residents). In reference to England, Raymond Williams (1973) argues that "the country" should here be understood as a relational concept whose significance is decipherable only in juxtaposition to "the city." The Mananara-Nord countryside was romanticized by town dwellers as an uncorrupt and idyllic place that could be tolerated for only a short while until either boredom set in or the vulgarity and childlike mentality of its inhabitants lost its charm. Peasants who lived in what townspeople would consider the *ambanivôlo* always located the outback at a more distant point from themselves. Villages that were more surrounded by primary forest than denuded mountains, for example, were deemed the true outback to villagers. It was rumored that in the nucleus of the remnant rain forest, a small and exceptionally primitive community subsisted, living in nature in the most rustic of dwellings, just as they had during the 1947 uprising of Betsimisaraka and other Malagasy against the colonial administration, when people fled into the forest to avoid being recruited by the growing rebel army or to escape the vengeance of the French (Cole 2001). But no one I met could ever confirm the existence of this community of "savages" (*sovazy*, from the French *sauvage*) with their own eyes.

The heartland of Mananara-Nord in the early 2000s was a patchwork of villages, agricultural plots of paddy rice and dry hill rice (called *jinja* or *tavy*), rain forest, fallow land with new-growth vegetation (*savoka*), and shady gardens (*ankôba*) of fruit trees, vanilla plants, and peppercorn vines. The deep *ambanivôlo* included 24,000 hectares of national park created by UNESCO, as well as the two hundred or so hamlets surrounding it. Villages consisted of houses built with plank and corrugated iron and ravinala bark and thatch. Most villages, including Varary, which became my primary field site because it was the residence of two ICDP conservation agents and a key access point for tourists heading into the core of the biosphere reserve, are accessible only by foot. With approximately 550 inhabitants in 2000, Varary was nestled in mountains that had been shorn of rain forest. Epitomizing the architectural and cultural traditions of northern Betsimisaraka society, Varary was a nucleated village within a hamlet "raised on poles and covered with split bamboo [or *falafa*] siding" and organized by bilateral kin groups. Newly married couples in the rural hamlets would live in the wife's village, husband's village, or mother's brother's village depending upon the availability of land and kin (Fanony and Wright 2003:54–55).

The Mananara-Nord Biosphere Reserve project began with a socioeconomic survey of the region conducted under the auspices of the Ministry of Animal Production and of Waters and Forests, UNESCO's Man and the Biosphere Program (MAB), and the Museum of Natural History in Paris in 1987 (Duvernoy 1987). The region was officially decreed a biosphere reserve in 1989, and the start-up phase of conservation and development lasted until 1991. The "transition phase" lasted from 1992 to 1995 and was funded by the Malagasy government, UNESCO, the United Nations Development Programme (UNDP), and the World Bank. Under phase 2, beginning in 1996, the biosphere's pilot ICDP proceeded under the joint funding of the Malagasy state and the Dutch government (through its international aid program, Directoraat-Generaal Internationale Samenwerking), while operated by UNESCO with the collaboration of the Association Nationale pour la Gestion des Aires Protégées (National Association for the Management of Protected Areas—ANGAP). The present and prospective strategies for management of the reserve also rely on collaborations with local associations and politicians of the district, including administrators of each municipality in the Mananara-Nord district.

Most of the coastal and montane villages of the district fell within the reserve's "peripheral zone," in which the ICDP implemented development activities. Many protected areas in Madagascar contain multiuse zones, in which residents may harvest natural resources so long as they obey rules that ban the killing and removal of endangered species (Nicoll and Langrand 1989; Orlove and Brush 1996). In this particular peripheral zone, the biosphere staff built a number of small dams, tree nurseries, and apiaries, and had trained residents in improved agricultural and fishing technologies (Brand 2000:24). Until December 2001, UNESCO and the Dutch and Malagasy states jointly administered both the reserve and the country's pilot ICDP, at which time ANGAP took over. ANGAP was the national park service, officially designated a "parastatal organization" by foreign donors to distinguish it from the Departement des Eaux et Forêts (the Department of Waters and Forests), the ineffectual public entity that governed forest concessions and parks. Financed by foreign aid and the revenue from the ticket entry fees to the nature reserves, ANGAP originally had the mission of ecological preservation, ecological research, environmental education, ecotourism development, and support for community development activities in peripheral zones of protected areas. ANGAP's name was changed in 2008 to Madagascar National Parks.

In the ICDP of Mananara-Nord, conservation agents comprised one component (*le volet*) of the ICDP, while their structural counterparts, the "development agents," were responsible for implementing poverty-alleviation measures, such as establishing tree nurseries, building dams for irrigation, and training villagers in beekeeping, sustainable fishing techniques, and child nutrition. These locally hired workers, whom the Malagasy call *zanatany* ("children of the land"), together accomplished a great deal during the early years of the ICDP, but their activities gradually stagnated. The development component of ICDPs was phased out once ANGAP took over a project from the international organization.

Plagued by budget deficits in the early 1990s, ANGAP opted to eliminate development activities. Development work was instead farmed out to a Washington-based nongovernmental organization (NGO), Chemonics, through the Land Development Interventions (LDI) program, which was in turn funded by the U.S. Agency for International Development (USAID). ANGAP's policy of phasing out community development activities to focus solely on ecotourism and conservation factored into my decision to focus on conservation agents; indeed, several of the former development agents of the biosphere's ICDP were retained by ANGAP as conservation agents.

The biosphere reserve's ICDP had tried for over a decade to win over rural villagers. The staff's attempts to get the peasants to adopt agroforestry practices and to respect conservation legislation met with sometimes fierce opposition. Local politicians of Mananara-Nord did not help matters. They tended to use anti-conservation sentiment to their advantage when campaigning in the countryside. Most residents of the Mananara-Nord prefecture identify ethnically as Betsimisaraka and practice *tavy,* so this population is a central object of conservation intervention and global anxiety about biodiversity loss. Like other coastal populations, Betsimisaraka (Malagasy nouns do not distinguish singular and plural) express intense animosity toward Merina, the politically dominant ethnic population. Hôva is the name that Betsimisaraka and other coastal populations (*côtiers*) call people of Merina descent, who claim the central highlands as their natal territory.[13] Betsimisaraka people's animosity toward Merina derives from the internal colonization of the island by the Merina king, Radama I (ruled 1810–1828). Seeking access to the eastern ports in order to trade slaves and goods, Radama declared his conquest of the east coast and of Betsimisaraka people (Cole 2003:102). In response to Radama's conquest, Ratsimalaho, the son of a princess from Fenerive and a New Yorker, appealed to the heads of numerous coastal lineages to establish what came to be known as the Betsimisaraka Confederation, a loosely organized group of lineages. The ethnonym *Betsimisaraka* is translated as "the many who will not be torn asunder."

King Radama I's conquest of two-thirds of the island and imposition of a caste system, slavery, and compulsory royal service sowed the seeds of interethnic hostility (Campbell 2005). This generally played out as a deep antagonism between the Merina of the central high plateaus and about sixteen other ethnic groups surrounding the plateaus and claiming coastal territories, collectively known as *côtiers*. As France set its imperial sights on Madagascar, Radama and his successor, Queen Ranavalona I, sought to enrich the Merina kingdom's coffers in order to defend against foreign invasion. Local Merina agents were deployed to curtail subsistence agriculture, to exact taxes and labor, and to capture slaves (Esoavelomandroso 1979:98). Betsimisaraka people today proclaim it taboo (*fady*) to marry Merina (Brown 1999). Ideologies of ethnic difference in Madagascar that took shape and solidified under Merina domination were later adopted and institutionalized through French colonial labor policies.

Europeans and Malagasy people of the central plateau region (the Merina and Betsileo populations) have disparaged the Betsimisaraka practice of *tavy* as wasteful and befitting of a people who loathe the exertion of building earthen terraces and irrigation canals on their ancestral lands (*tanindrazaña*). Missionary and ethnographer Paul Cotte judged Betsimisaraka to be generally "without pity" toward the forest, called "the robe of the house of the ancestors" (*simbon'tranony razana*) by Malagasy (Cotte 1946:6). In contrast, outsiders have admired the practice of paddy riziculture in the central plateau region, where the farmed landscape conjures the image of East Asia because of its vibrant green, terraced slopes.

Merina individuals have occupied most of the high-level positions and ICDP directorships in the conservation and development bureaucracy, and the politics of ethnicity, as well as peasant resistance to conservation, contributed to the ICDP's rocky start. The politics of ethnicity among Malagasy groups is something that many expatriate representatives of conservation and development discover only once they move out of the planning rooms of agencies in Antananarivo and into the sites of protected areas on or near the coasts.

THE CHAPTERS

The book places present-day forest conservation in Madagascar into the context of forest-based activities occurring over the past century. The larger canvas of time and space gives a better sense of how values of nature in Madagascar have emerged and evolved through the movements of people over the land, and through interactions of people from different social strata and cultural milieus. The stories told in each chapter do not follow a strict chronological progression. Following my informants, who tended to recall the colonial period as they negotiated the intrusions of conservation and development in the early 2000s, I put the period of my research in dialogue with the colonial period. In general, however, the first three chapters are anchored more firmly in the late nineteenth and early twentieth centuries, while chapters 4 through 7 emphasize the ethnographic material collected between 2000 and 2002.

Chapter 2 examines the semiotic representations of hierarchy in Madagascar established within the imperial caste system of Imerina, the earlier name for Merina territory, which provided a structural template for the French state. By way of social metaphors of feet and eyes, walking and vision, I discuss how the social organization of Madagascar informed how different groups have perceived nature, particularly the eastern rain forest. The kinds of knowledge born of ambulatory and optic regimes of perception have shaped the division of labor for contemporary conservation and development.

Chapter 3 addresses labor structure and internally opposed work ethics in Madagascar by comparing the value systems of agrarian and capitalist production in the eastern rain forest. The French sought to bring the rain forest of Mananara-Nord into capitalist production by commodifying Betsimisaraka labor, a process that began with the institution of compulsory and wage labor in logging, mining, and cash crop enterprises. Rather than focusing on instances of peasant resistance to capitalism and compulsory labor, the chapter analyzes how Betsimisaraka values have intruded into capitalist work sites, pushing against the temporal and spatial parameters of their production processes. The paradoxical relationship of the capitalist ethic and the subsistence ethic in Mananara-Nord centers on the productive aspects of their friction, as well as the forms of value they generate.

The colonial forest service in Madagascar carried out the first systematized conservation efforts on the island. Echoes of its bureaucratic structure resonate in contemporary conservation institutions in Madagascar. Chapter 4 examines institutional labor relations and the discourses of value concerning forestry and conservation, attending to the changing criteria of value for foresters and conservation workers in the colonial, socialist, and neoliberal eras. The chapter reflects on the implications of privileging intellectual labor for the purpose of rain forest conservation and examines the means by which subaltern workers have sought to advance their careers by assimilating the values of structural superiors.

As imperialist interventions of the late twentieth century, conservation and development efforts do not evacuate other ways of producing value but instead compete with preexisting modes. Chapter 5 examines the significance of clove and vanilla production in Mananara-Nord for biodiversity conservation and the production of rural conservationism. It examines the forms of sociality intrinsic to the cultivation, processing, and exchange of cash crops. It also considers how the social relations of cash cropping, involving duplicity and intimidation by representatives of the state and export firms, can work to undermine conservation efforts even though cash crop production appears to complement it by luring peasant producers into the global market economy and thereby reducing pressure on the rain forest.

The global conservation effort lays emphasis on getting people to appreciate the intrinsic goodness of endemic species and primary habitats, and a key way in which nature's intrinsic value is constructed in poor countries is through the concept "world heritage," which is in turn broken down into units of natural and cultural heritage. The selection of heritage sites, as well as the publicity afforded to certain species, landscapes, and cultural formations by organizations such as UNESCO, serves to define group identity at nested scales: global (humanity's common heritage), national (Malagasy heritage), and regional identity (various constellations of ethnic heritage). Not only are these identities imposed from on high, but they also filter out cultural and biological forms that may be of utmost importance to social groups yet undermine the mandates of neoliberal development. Chapter 6 explores heritage as an operative concept in Mananara-Nord, and takes "ancestral custom" (*fombandrazaña*) to be the obvious analog. It describes how conservation agents negotiated their duties to restore and protect the natural-cultural heritage privileged by UNESCO, and their duties to fulfill their obligations to dead ancestors, which entailed specific ways of exploiting the land and interacting with organic beings. While the respect for ancestral custom, which established how to exploit land and interact with natural entities, maintained a sense of group cohesion and historical identity for Betsimisaraka residents, including conservation agents, the reproduction of *fombandrazaña* often directly thwarted the ICDP's effort to valorize natural heritage.

Chapter 7 considers the subjective experience of structural contradiction, or how manual ICDP workers negotiated their work lives as peasants and conservation implementers. The chapter focuses on two phases of the ICDP in which conservation agents were demoralized and felt a high level of anxiety. These two episodes included the forest sweep, where conservation agents allied with gendarmes to rout out from the national park peasants who had illegally cleared land there, and the period of management transfer when UNESCO, and the Dutch and Malagasy states, prepared to hand over control of the ICDP to the national park service, ANGAP. The liminal period of late 2001 intensified the frictions between neoliberalism architects, the postcolonial Malagasy state, and non-elite, rural people. The latter not only were pulled into and affected by the valorization projects of outsiders entering their lands, as well as by the tactics of state officials trying to keep their positions of privilege, but also suffered the accelerating depletion of endemic species and habitats. The dual modes of production in Mananara-Nord, and elsewhere in the global South, have produced a market in which "culture," and specifically the culture of labor, should exist in the service of an extinguishing nature in order to valorize itself, as though the rural poor have been primarily responsible for the eruption of biodiversity hot spots on Earth.

“A forest is a gold mine to a Naturalist.”
The live hush—rustle—reverence.
He feels walls of his life dropping away.

You need people who know
  the broken trails, sudden pits underfoot,
and the animals. Capybara, jaguar, agouti.

—RUTH PADEL, “In the Seraglio,” *Darwin*

2 Overland on Foot, Aloft

An Anatomy of the Social Structure

A European traveler to Madagascar in the early nineteenth century, say 1825, would encounter a mosaic of rolling grassland and humid rain forest outside the limits of the eastern port of Toamasina. The traveler would likely head westward to pay his respects to King Radama I and his court in Antananarivo, the seat of the Merina Empire on the central high plateau. The trek from the coast to the capital was over 200 miles long, and the traveler, possessing heavy trunks of clothing and food provisions, faced an uphill and uncomfortable journey through a rain forest that, for all of its botanical and zoological wonders, could be lethal. Malaria had felled many. It was said that King Radama's military strategy relied on "General Hazo" and "General Tazo" (Generals "Forest" and "Fever") to shelter the Merina kingdom from foreign invaders (Gallieni 1908:149; Campbell 2005:245).

Madagascar's east coast had heavily trafficked ports because of the relatively calm waters of its harbors. Toamasina in particular was reputed to offer the best anchorage of the island (Lloyd 1850:59). The east coast was thick with precious timbers, minerals, and fruits, and it possessed a well-trafficked footpath between Toamasina and the highland capital, Antananarivo. Automobiles would not appear on the island until 1900, four years after France's annexation of Madagascar and two years after Governor-General Gallieni actually purchased the cars from abroad—two Panhard-Levassors (Gruss 1902:194). In 1900, a celebrated "road to the east" from Antananarivo was opened (Gallieni 1908:170). The Tananarive-Côte-Est (TCE) railway would not be completed until 1913, built with the exertions of Malagasy laborers who were drafted by the French state during a huge public works campaign (Gallieni 1908:226; Porter 1940).

The path to the High Plateaus was often treacherous and steep, crosscut by rivers and root systems. The traveler had no choice but to go overland, carried in the *filanjana,* the Malagasy palanquin. This was a raised chair hoisted on the shoulders of four to eight Malagasy carriers, tradesmen known as *mpilanja* (alternatively, *mpilanjana*) (Valmy 1959; Campbell 2005:252). On the *filanjana,* the European's legs and feet were Malagasy. Europeans and Malagasy together subscribed to the idea that the forested footpaths were unsuitable to the European constitution. In Max Mezger's 1931 German children's novel, *Monika Fährt Nach Madagaskar* (Monica Goes to Madagascar), a scientist father explains to his young daughter upon their arrival on the island why they must travel by *filanjana* instead of walk: "Only the natives with their broad feet and soles as hard as leather can walk fast on such ground" (Mezger 1936:148).

This is to say that before the appearance of bush taxis or locomotives, European visitors encountered the eastern rain forest without a direct connection to the ground. At a minimum, they wore footwear; most often they were transported in *filanjana* and canoes. Lurching along in their elevated chairs, Europeans surveyed the environment, their hands free to sketch and record observations in their journals (see Pratt 1992). Many returned to England and France to publish travel journals illustrated with exotic plants, animals, and the island's diverse "tribes." These natural histories were the products of contemplation unmoored from the weight of the body while walking. Like travel, the act of translating a landscape into words and images defined a modality of perception that was specific to the time and place (Raffles 2002). Botanical exploration in Madagascar, as in early-twentieth-century Burma, Tibet, and China, entailed "practices of writing, revising, and reading the landscape to fashion Edens of the world of things" (Mueggler 2005:447). For the orchid hunters, ethnologists, and natural historians who traversed Madagascar's rain forest by *filanjana,* such a mode of transport may not have been ideal, but it enabled a slow-motion translation of nature into text. It also placed the traveler on a stratum of space that was free of the sensory intrusions of walking, such as thorny scrub and root-gnarled ground, narrow corridors of hardened clay, and blood-sucking leeches that numb the skin then, bloated, drop off, leaving itchy welts.

Yet sitting in a *filanjana* could be acutely uncomfortable. As Reverend William Ellis, plant collector for the Royal Gardens at Kew and member of the London Missionary Society, describes in his 1850 account of *filanjana* travel: "The clayey sides and rocky portions of the ravines were sometimes so steep that my position was almost upright, and it frequently required ten or twelve men to get the palanquin up and down" (Ellis 1859:319). At other moments, however, the passenger of the *filanjana* was well positioned to drink in the scenery while the mind cogitated. The difference in station between the traveler and the carrier, made manifest by their physical positioning in space, influenced how each perceived wildlife and human society. This was not an essential difference of sensory faculties, but rather the result of a political and cultural history that reverberates into the present day in Madagascar and informs the program of biodiversity conservation and park tourism.

Contemporary tourists and scientific researchers now travel in more modern forms of transportation, but modalities of perception are always constituted by social structure and inform and justify the division of labor. For the business of protecting species and habitats, the "intellectual labor" of foreign experts—some who stay a short while in Madagascar, and others who make the island their permanent home—has overshadowed the contributions of subaltern labor. This category consists of Malagasy porters, builders, servants, and guides, who for a small wage have made life livable for foreign travelers and who have enabled the conditions of possibility for scientists to discover new species in Madagascar's ecosystems.

In this chapter, I focus on a constellation of signs that refers to how particular actors have valued nature in Madagascar—and by nature, again, I mean both the perceived inner quality of human beings and the outer form of biophysical objects. I narrow down the semiotic field to two key symbols, feet and eyes, and two historical modalities of transport, walking and being carried. As I will explain, the feet-eye symbolism was a recurrent theme among Betsimisaraka residents of the biosphere reserve during the time of my ethnographic research. It came up when I was not looking for it, and then I began to attend to its appearance. But it was only on later reflection that I saw a correspondence between the anatomical symbolism and a wilderness epistemology that was tied to a person's social position and the extent to which that determined his or her modality of travel. Eyes and feet are bodily symbols of social hierarchy as well as sensory apparatuses attuned by class structure and the process of labor.

OPTIC AND AMBULATORY REGIMES OF PERCEPTION

Writing about the social history and phenomenology of walking in Britain and Europe, Tim Ingold (2004) argues that the mechanized manufacture of footwear around the eighteenth century was part and parcel of historical transformations in

> modalities of travel and transport, in the education of posture and gesture, in the evaluation of the senses, and in the architecture of the built environment—all of which conspired to lend practical and experiential weight to an imagined separation between the activities of a mind at rest and a body in transit, between cognition and locomotion, and between the space of social and cultural life and the ground upon which that life is materially enacted. (Ingold 2004:321)

Translocating Ingold's analysis to the context of colonial Africa, where Europeans were exploring territories and plotting settlements, changes the representational meanings of ways of getting around. In a colonial context, outward expressions of European class, such as the type of shoe worn, for example, or a certain posture and gesture, or the ability to take a leisurely stroll versus the required drudge of trekking to one's field to farm, are interpreted by African others according to a completely different matrix of value.

Nineteenth- and early-twentieth-century accounts of travel through African landscapes verify Ingold's claim about the affluent of Europe who "came to conduct and write about their travels as if they had no legs" (Ingold 2004:322). Charles Darwin remarked on the relatively prehensile foot of the unshod "savage" compared to that of "civilized" man (Darwin, 1874:77, cited in Ingold 2004:318–319). I noticed this physical trait myself in the hamlets of eastern Madagascar, where the big toes of many Betsimisaraka peasants curl outward, trained from a lifetime of being inserted like crampons into the steep, slick trails of the eastern scarp. I expect that the shoeless feet of nineteenth-century *mpilanja* were also bowed and leathered from their trade. European travelers had to trust the surefootedness and "ambulatory knowledge" of their Malagasy carriers.

Feet and eyes are key symbols of the relational positions of Malagasy and *vazaha* (foreigners, strangers). This was brought to my awareness in March 2001 as I watched a newscast in Antananarivo at the home of my friend and former research assistant, Haja, the young Merina woman who was my assistant during my first fieldwork stay in Madagascar (1994–1995). In March 2001, the Malagasy public was gripped by two imminent events. First, the public was waiting for then president Didier Ratsiraka to announce whether he would run for reelection against the Merina yogurt-business magnate, Marc Ravalomanana. It was the first time that anyone had posed a serious threat to Ratsiraka's rule, and the political machinery of his long-entrenched party, AREMA, and Ravalomanana's new one, Tiako Madagasikara ("I Love Madagascar," a pun on his dairy company's name, Tiko), were quietly preparing for a fierce campaign battle.

The second imminent event was a total solar eclipse scheduled to occur in June. Southern Madagascar was supposed to offer the best view, so hotel rooms in the south of the island had been booked by foreign tourists for months ahead of time. Although the coming eclipse was an economic boon, its approach induced anxiety among Malagasy citizens who believed it would cause blindness, destroy crops, and create mayhem. Christian Malagasy, on the other hand, welcomed the coming eclipse as a sign of the Apocalypse.

In the newscast, Ratsiraka asserted that he did not want the "election to eclipse the urgent matter of the solar eclipse." As proof of his concern, he announced his wish that all 15 million citizens should receive special cardboard and plastic-film glasses to protect their eyes from the sun's rays, and he invited on the air a famous savant-diviner (*ombiasy*) to interpret the eclipse to television viewers. The *ombiasy* wore a friar's robe, a pendant around his forehead, and had long, braided hair. He began to expound on the symbolic and spiritual meanings of the eclipse—I could glean only parts of his lecture in Malagasy. He sketched a map of a man standing upright, back turned to the viewer, arms raised. Across the back of the man's spine, the savant drew a horizontal line, and wrote beside it, EQUATOR. He likened the male figure to the globe, where latitudes and longitudes corresponded to the human body. The dimensions of man's feet were proportional to Madagascar, he explained. The future of Madagascar lay in the man's palms, France. Hands and fingers were somehow analogous to phases of the moon (I could not easily follow him), the symbol of mother for Malagasy, while the sun symbolizes father, he explained. The eclipse signified their "cosmic, conjugal union." A sociologist in the television audience asked the savant whether "the occidental technological knowledge" that claims it is dangerous to gaze directly at the eclipse is linked to the occidentalist prohibition against gazing upon Mother and Father in the act of consummating their union, to which the savant nodded.

In the *ombiasy*'s map, the head and eyes represented "occidentalist technological knowledge," a privileging of intellectual discovery and structural superiority, as represented by the man-shaped landmass with uplifted arms, with France in the palms. Madagascar, the feet, was the unanalyzed counterpart—by implication Madagascar lay at the foot of Empire. One could make a persuasive argument that the feet in the *ombiasy*'s illustration represented the subaltern subject, subjugation, and the foundation of wealth production.

Beyond the point that the Merina or European colonizer has ruled by way of vision-centric technologies—curiously, during the French invasion of Antananarivo in 1895, the French army targeted a French-built meteorological observatory that the Merina queen had shut down and turned into a fortress (Colin 1898:308)—feet symbolize unequal power relations that predate European-Malagasy contact in Madagascar. Members of royal Merina lineage used to demand of their subjects gestures of self-abjection, such as foot kissing. The phrase *milela-paladia*, kissing the sole of the foot, signifies abject submission, and in Madagascar's past was an act of respect from slaves to masters, wives to husbands, and inferiors to superiors (Sibree 1884:177). The phrase persists in the present day.

ORIGINS OF THE *FILANJANA*

The *mpilanja* trade developed under the reign of King Radama I, but the history of the *filanjana* dates to the eighteenth century, coincidentally the period when the mechanization of footwear manufacturing took off in Europe and for elites the necessity of walking was alleviated by advances in transportation (Ingold 2004). Members of the royal Merina family were carried by slaves in *filanjana* with leather seats. The royal *filanjana,* unlike other regional varieties, was quite heavy and required two teams of four men, which took turns carrying it at five-minute intervals, according to French geologist and historian Robert Valmy (1959). Max Mezger (1936:148) describes how the second team of *mpilanja* could relieve the first team while en route, without breaking the stride. Gradually *filanjana* travel became accessible to other elites, as in 1821 Radama formally lifted the ban that limited the use of palanquins to court members (Campbell 2005:252).

By the late 1800s, the portage system within the Merina realm included as many as 60,000 slaves (Campbell 2005:253). Gwyn Campbell estimates that these "slaves formed from 19 to 23 percent of the entire slave population of Imerina" (Campbell 2005:253) . Although they were slaves, porters received an important share of foreign exchange. The income benefited the slave syndicates and strengthened their hold on the portage trade. Campbell writes that porters as a group were well organized and unified, a proto–trade union: "Despite their slave status, porters were the only large wage-earning body of workers in imperial Madagascar, lived a more independent lifestyle than free subjects, and behaved in many respects more like wage earners in Europe than did other Malagasy workers—most of whom were, under the imperial regime, subject to *fanompoana*" (Campbell 2005:256). *Filanjana* portage earned a minimal wage (*adimbilany,* "enough to fill a pot"). *Mpilanja* worked closely with the *maromita* or *borizano,* the porters of cargo and luggage.

They were highly skilled and developed different gaits for transporting their passengers. The smoothest and ideal type of gait was called *rano mandry* ("sleeping water"). But should the passenger prove difficult, or should the carriers want to amuse themselves, they had recourse to a variety of gaits. Valmy lists them:

> *Betsimisay* or *mitofitofina:* a kind of unequal step which resulted in passenger fatigue.
>
> *Manakalahy:* an intermediary gait between sleeping water and racing.

> *Bakose:* consisting of elongating the stride and advancing with outstretched legs.
>
> *Banansandrata:* a disagreeable gait resembling a limping horse that gave the impression that one was going to crash and fall with each step. (Valmy 1959:41)

Colonial scientist Alfred Grandidier noted one of the songs of his company of carriers: "Look master look! See how I am strong! See I don't have fever, my feet are not pained! Look well, master!" Valmy also remembers the constant chatter and "exuberant temperament" of his carriers on one of his voyages. He quotes one as saying, "it is the whites who have the shoes to protect their feet, and they who carry them go barefoot!" (Valmy 1959:41). Despite their literal position as underlings, *mpilanja* wielded some power. They could strongly affect the passenger's travel experience by controlling the smoothness and length of a day's journey as well as the level of the ambient noise, such as chatting and song. All of this made an impact on the traveler's bodily awareness and affected the context in which he or she wrote, sketched, or gazed down upon the natives "ravaging" the forest by slash-and-burn.

In the present day, Malagasy people remember the work of portage during the colonial era. Betsimisaraka villagers in Mananara-Nord have distinct memories of carrying French officials in their palanquins. A man named Rasonina, for example, was a rich source of oral history. He was the father of a biosphere reserve conservation agent and the *tangalamena* of Varary village, where I spent about a week or two every month doing research. The *tangalamena* (literally "red baton") is a Betsimisaraka spiritual leader who presides over rituals that commemorate ancestors (Fanony and Wright 2003). Rasonina was in his late sixties and considered the village historian. He recalled what life was like during "the time of the *vazaha*." He likened those days to slavery because the *vazaha* "were carried around to go places. When they were on your shoulders, there were some who felt like peeing. From above they'd pee on the shoulders—they peed on the shoulders of men! It's true" (Field notes 3/10/2001). While Rasonina acknowledged that some *vazaha* were "kind-hearted to Malagasy people" and would descend from their elevated seats, the memory of *vazaha* "peeing on the shoulders of men" troubled him. Rasonina repeated it three more times, apparently awed by this act of debasement by *vazaha*. Jennifer Cole (2001:165) similarly writes about the recurrent memory of palanquin-carrying among Southern Betsimisaraka people of Mahanoro: "Countless people—both those who had lived through colonial times and those who were born well after it was over—told me about the palanquin. They never failed to mention the fact that people riding in them urinated and defecated on those below." Seeing pictures of certain models of the *filanjana*, the claim seems doubtful, the position contortionist. In any case, the memory may speak of the truth of experience. It expresses the historicity of colonialism, the subjective experience of subjugation and its enduring scars, even if the factual accuracy is unclear.

ISLAND VISIONS

Philosophers and linguists argue that metaphor operates at a cognitive level. It guides the development of language by establishing relevance between two things that are not immediately or obviously connected (Lakoff and Johnson 1980). Inspired by this claim, Anne Salmond (1983) dissects a pervasive metaphor in the English language—"knowledge is a landscape"—and reveals how this metaphor links up to others to forge a metaphorical chain: "Intellectual activity is a journey" through a landscape; "destinations" and "facts" are like "natural objects." We must conclude, she asserts, that seeing is equal to understanding (Salmond 1983:67). In Salmond's analysis, the dark space of the unknown is illuminated as knowledge progresses forward in space, as it is possessed, mapped, and charted (Salmond 1983:69). Metaphors are the building blocks of a spatializing and hierarchizing "language of value" (Ferry 2005:17). They establish sight as the privileged sense, the progressive sense that has been inextricably bound to imperialist conquest.

Contemporary Western imagery of Madagascar as a wounded, dying island—an image frequently invoked in conservation and ecotourism appeals—came into being through the imperialist encounter. The early nineteenth century marks the period in which European missionaries, colonists, scientists, and adventurers began to visit Madagascar's interior in greater numbers. Because the European traveler had no choice but to be delivered via the island's portage system, he or she—including the renowned Austrian explorer Ida Pfeiffer, who succumbed to disease in 1858 not long after her trip to Madagascar—not only arrived with whatever optical and writing instruments were available to amateur natural historians, but was also ensconced within the particular stratum of space and social structure mediated through the *filanjana.*

Several books and articles describe how Europeans experienced this mode of travel in Madagascar. Reverend Richard Baron, commissioned by an Antananarivo-based committee of the London Missionary Society to visit churches and schools in the northern regions, made an excursion in 1891 and published his account the following year as "Twelve Hundred Miles in a Palanquin." As he travels by *filanjana,* Baron decried the "terrible havoc" he witnessed in the eastern forest, caused by "axe and fire." His observation of *tavy* while en route, seated above the scorched earth, provoked him to cogitate on deforestation on a larger spatial and temporal scale. He saw in his mind's eye the forests of the eastern slopes beset on two fronts by *tavy,* and ultimately gone in a not-too-distant future:

> The fact is, the forest on the east side of the island is not merely being thinned (which in itself would be no evil) but is being absolutely laid waste by the natives, who, like two lines of despoilers, stretching, one on the east and one on the west side of it, for many hundreds of miles, commit their depredations the whole year through. Probably more than one half of the original forest has been already cut or burned down, and in a few generations, at the present rate of destruction, this still magnificent mass of vegetation will have been swept out of existence. (Baron 1892:434)

In 1928, Charles F. Swingle, a twenty-nine-year-old botanist with the United States Department of Agriculture, visited Madagascar accompanied by Professor Henri Humbert, a renowned French botanist, in quest of a species of rubber plant believed to have been "exterminated" by overexploitation. Swingle traveled overland in a *filanjana* as well as by "boat, auto, and railroad." In a *National Geographic* article about his trip published after his return, Swingle waxes nostalgic on the "desolation of the mountains" after their denudation. He states that the "hills, once covered by splendid forests, were now brown and barren save where grass had sprung up, as if attempting to hide the all-too-evident traces of the destruction wrought by man" (Swingle 1929:179).

In the past as in the present, dire prognostications and assessments of Madagascar's environment were tempered by appreciation of its scenic beauty. In a report to the Royal Geographical Society of London, T. Wilkinson described a voyage from Toamasina northward to the island of Saint Marie. He remarks on the "conspicuous orchids" and "park-like" and "romantic" scenery of the coastline (Wilkinson 1869–1870:372–374). My quotations of Europeans' accounts are not intended to suggest that nineteenth-century Malagasy did not also appreciate a good view. For a Malagasy peasant, however, a pleasant forest scene evoked something different from parks or biblical gardens. On the now denuded spine of the eastern escarpment, the Zafimaniry people, writes Maurice Bloch (1995), appreciated the clear and spacious views unencumbered by dense forest and mist. In the past, slaves were required to live in accessible, low-lying villages. Once freed, former slaves ascended to mountain summits to settle. For the Zafimaniry, the desire to overcome the obscurity of forest was thus bound to the experience of forced labor and eventual freedom (Bloch 1995).

Europeans' travel accounts of Madagascar melded their interpretations of Malagasy nature with Malagasy character. Swingle devoted quite a lot of text to his porters, who were employed to convey him for a total of 400 miles from Toamasina to the southern desert region. He described their fortitude and laudable perseverance during a stretch through a desert with an insufficient water supply as well as their lack of "grumbling" when carrying their loads. But at the task of digging up specimens of *Euphorbia intisy*, the rubber-bearing plant thought to have gone extinct, the porters balked: "they were soon showing me blisters on their hands," he writes, "and trying to persuade me that the job was out of the question" (Swingle 1929:202–203). But since Swingle was determined to bring back the plant to Washington and the botanical gardens of Antananarivo, Algiers, and Marseille, he paid each porter "the unheard-of wage of 40 cents for his day's work" to obtain the specimens he desired. In sum, Single judged the porters "were a good-natured lot," although it was often necessary to "reprove them for singing and laughing all night long" (Swingle 1929:202–203).

Singing and laughing as they carried passengers by day, while undoubtedly making work more tolerable for porters, impinged on the sensibilities of the foreign traveler awkwardly perched in the "precarious seat" (Swingle 1929:195). An 1883 dispatch by a *London Standard* correspondent, copied in a *New York Times* report the same year, reflects on how an Englishman's assessment of the Malagasy forest enfolded Malagasy society and wilderness:

> The philanzan men paddled the canoes. Of the singing of the men [the London correspondent] says: "There were bass singers and seconds, while the melodies were often sweet and strangely plaintive. Indeed, we fancied we could trace in this Malagasy music indications of the higher civilization of which the people are capable. . . . A Malagasy path goes very straight. It despises easy inclines and revels in precipices. Ant-like, the Malagasy goes over everything in front of him. He prefers, if anything, the highest point in any range of hills that may have to be crossed, and deviating neither to the right nor to the left, goes straight over it, no matter how steep the slope may be. Our daily toil for nearly a week was the continual ascent and descent of rocky ridges covered with forests on the tops, which, running in seeming endless succession due north and south, and at right angles to the path, render penetration to the interior of Madagascar from the east coast so difficult. . . . Sitting in a philanzan during such a journey is no very pleasant experience. A solemn and oppressive silence prevailed everywhere. There is no life in the woods of Madagascar, and throughout our journey a single black parrot and a melancholy crow or two were the only birds we saw." (*New York Times* 1883:2)

The jerking movement of his body suspended above the ground, raised above his carriers yet at their mercy, intruded on his experience. He deemed Malagasy to be uncivilized, "ant-like" makers of despicable paths, and the Malagasy rain forest to be oppressive and melancholy. Was the stillness of the Malagasy forest due to the paucity of life in it, or was it an effect of its relatively quiet creatures?

The hegemony of mediated sight for Europeans—sight mediated by the watchful eyes of Malagasy carriers, by satellite images and photographs, and by pencil and paper and camera lens—has obscured the significance and very existence of grounded labor and knowledge. Botanical and zoological explorers in the tropics relied on an entourage of local porters and guides whose labors facilitated scientific knowledge, yet whose insights were eclipsed by it. The European perception of the Malagasy landscape and the Malagasy peasants' actions upon it resulted in the legislation of specific temporal geographies: nature reserves as snapshots in time of rain forest space, plucked up and protected from the hubbub of production. Moreover, the process of species discovery has led to closer examination by scientists and conservationists of areas with endemic biodiversity and erosive land use practices, enabling scientists to determine whether habitant species should be included on the Red-List of Endangered Species compiled and published by the International Union for the Conservation of Nature (IUCN).

A colonizing gaze has served to construct environmental crisis in Madagascar. Gillian Feeley-Harnik (2001:32) has compiled Western visual and verbal imagery that depicts the island in its death throes. This discursive turn has been particularly evident since the 1980s, when the island achieved "hot spot" status and an influx of foreign aid. Feeley-Harnik offers several examples, which I sample below:

> a. The author of "Island of Ghosts" writes: Madagascar is "vanishing before our eyes, being consumed from within . . . burning to death."
>
> b. French zoologist and botanist Jacques Millot (1972:752) writes in 1972: "Destructive human activities sometimes create spectacles of diabolical beauty. The Betsiboka in flood can be seen to tear from its denuded banks so much red earth that its waters become as though stained with blood."
>
> c. Jacques Hannebique's (1987) popular photo-essay, *Madagascar: Monîle-au-bout-du-monde,* offers an image of Madagascar from the air: the channels of the Betsiboka River spreading out below a plane heading to Antananarivo: "At 10,000 meters of altitude, in the plane which, from Europe, takes you toward Madagascar, well before seeing its shores, you are suddenly intrigued by a immense reddish spot, which stands out sharply against the blue water of the Mozambique, without mingling in it . . . That red water, spurting out from everywhere like a fatal hemorrhage whose power forces the sea waves beyond the horizon, is *the blood of the earth.*" (ibid.:8, 9, emphasis in original)

A late addition to these images is a 2004 space station photo, published in *National Geographic,* of the Betsiboka estuary in the wake of the Cyclone Gafilo, revealing plumes of red sediment branching off the river. The image substantiates the French and North American trope of Madagascar's "red earth bleeding into the sea . . . from poor agricultural practices" (Hannebique 1987 cited in Feeley-Harnik 2001:32).

THE EXPRESSION OF AMBULATORY VALUE

As a primary mode of travel in rural Madagascar, walking continues to be a marker of one's social rank and ability to make a living. In Malagasy, the word *hôngotra* (or *tongotra* in Merina) means both "foot" and "leg," and the phrase *mandeha hôngotra zahe* means "I'm going on foot." Whether one planned to go on foot, or the fact that one was presently on foot, is a ubiquitous bit of small talk in Mananara-Nord today. People also talk about the condition of roads, the distance of locations, and the speed of pedestrian travel.

The importance of walking, and therefore feet and legs, is especially evident in funerary practices. In 2001 I was attending a ritual in Varary in which two men's bones were being repatriated to their ancestral tomb from a faraway village. (In Varary extended families possessed familial tomb sites [*ala fady*] shrouded by primary rain forest because it is *fady* [taboo] to raze the forest of the dead. To transgress this *fady* risks angering the ancestors, who have the power to inflict death, harm, and other hardships on their living kin.) I followed a large group, perhaps half the village, to a hilly and shaded remnant of forest about a mile from the village. There, I saw coffins of the newly dead lying exposed on the tops of large boulders and covered in bright sheets of royal and powder blue plastic. I was told that they would remain like that, unburied, until the flesh decomposed inside the coffin, after which the bones would be removed from the coffin, wrapped, and placed in a cave (*lava-bato*), if one existed, or else a dugout hole, its opening sealed with a wall of stones. Clans interred all their dead males in one *lava-bato,* females in another. Sex segregation deterred incest among the dead, people explained. At a distance from the primary familial tomb lay sex-segregated tombs for kin who were diseased in life and who could potentially contaminate the other ancestors.

But when I reread my notes later, I noticed that only three kinds of illness required separate burial: leprosy (*bôka*), polio (*zaza-vatana,* or "child-body"), and lameness (*malemy hôngotra,* or "limp foot/leg"). Contagion alone did not qualify a body for separate burial: leprosy and polio are contagious, but so are cholera and tuberculosis. Lameness is not. The common denominator of the three illnesses involved a disability of the legs and feet, so the distribution of bodies in the space of tombs ranks them according to their value in life as able-bodied walkers, and therefore workers.

The organization of the tombs, in which the "healthy" dead were separated from the disabled dead, was also echoed in the classification system for spirits. Varary villagers recognized diverse spirit types. In order of most humanlike to least, these included ancestors (*razaña*), who were honored; ghosts (*angatra*), who were feared; *tromba*, possessing spirits that were usually dead ancestors or sovereigns; *kalañoro*, heritable spirits who served human owners as spirit mediums; and *tsiñy*, wild spirits who inhabited natural features, such as waterfalls and deep pools, and who might cause harm. *Kalañoro* were described to me as tiny, hairy, and fanged people who used to inhabit the rain forests until habitat erosion caused them to die out, only to return in spirit form. They have backward feet, which makes it very difficult to track their steps if they steal from you. Lesley Sharp, who carried out ethnographic research in the northwest Sambirano Valley, reported something similar. The people there described them as "dwarfed and hirsute, with unkempt hair, long fingernails, eyes that glow red, and feet that face backwards on their short little legs. . . . If captured, a kalanoro's human guardian can draw on this spirit's power to heal" (Sharp 2001:56).

Like slaves, *kalañoro* were said to faithfully serve their owners (*tômpony*) as spirit mediums, helping their owners set themselves up as clairvoyants for paying customers, and often pilfering things, such as money and beer, on their behalf. No one actually sees *kalañoro*. People who seek the clairvoyant services of *kalañoro* owners visit the owner's house at night. The owner typically speaks to the customer from behind a wall or curtain at night, the room dimly lit by kerosene lamp. The owner's voice becomes nasal and high-pitched, a sign of the *kalañoro* speaking through him.

Angatra (ghosts), in contrast, look like regular people and can be recognized only by their behavior and one anatomical difference: they have no feet. These secretive spirits come out only at night, and their legs, which end at the ankles, hover above the ground. If one encounters a ghost, I was told, one must attack it or else bad things may happen. Of the spirits, then, the two types described as having something remarkable about their feet, both of which seem to have an aversion to being *seen*, happen also to be the two types that are subservient to or reviled by living people because they can cause harm. As the "realm of the dead," the rain forest shrouds a cultural value that reveals itself only through the spatial representation of dead bodies (Feeley-Harnik 1984:12).

TOURING THE NATIONAL PARK

All this business about the importance of walking ability in rural Betsimisaraka society, as well as the symbolic dichotomy of eyes and feet, has a resonance in the sphere of conservation and development today. The distinctive and interdependent industries of conservation and ecotourism in the island's national parks rely on manual laborers to guide tourists through the rain forest, and these tourists often have a very hard time of it. The relationship between European or North American ecotourist and Betsimisaraka guide reflects much of the same complex arrangement, where the tourist is completely dependent—sometimes helpless—and the Malagasy guide is aware of his or her superior strength and agility yet also of his subordinate position and relative lack of worldliness (usually expressed to me in terms of misgivings about being incompetent in foreign language). The guide wants to please the foreigner for possible rewards (tips or gifts) and appreciates the experience with the foreigner because it develops the guide's skills and raises his or her level of sophistication (that is, the ability to interact comfortably with foreigners), yet the guide also feels the burden of having to keep the tourist unharmed and satisfied. This can entail carrying the tourist's heavy packs and provisions (and sometimes their bodies), building makeshift shelters in the forest if the rains come down, and cooking for the tourist during overnight camping.

Varary village was a critical base camp for tourists wishing to enter the biosphere reserve's rain forest because it lies only 6 kilometers from the drivable road and only a two-hour hike into the reserve's interior (a national park). For this reason, Varary residents were more accustomed to seeing *vazaha* than were residents of other villages.

Because of Varary's geographical position and the fact that most of its population had been vocally opposed to the park's creation, it had two resident conservation agents: Sylvestre and Jafa. Sylvestre, who was in charge of tourist-related activities, spoke a little French. He was the eldest son of Rasonina, the *tangalamena* of Varary mentioned above. As the eldest son, Sylvestre would inherit his father's role upon his death. The *tangalamena*'s main rival at the time, whom I call Lema, was the father of Jafa and one of the main detractors of the conservation effort in Varary. I will return to Jafa's life in a later chapter. Sylvestre had worked for the ICDP since 1987, when the initial sociological surveys of the region were conducted under the auspices of Eaux et Forêts (the forest service), the Worldwide Fund for Nature, and the former minister of education for President Albert Zafy. During the time of my research, he was also the *Parc Responsible,* the head of the park in Varary. At forty-seven, he was father to seven children, ranging in age from seven to twenty-seven, and grandfather to nine children.

The conservation agents were delegated the responsibility of tourist guidance only in 2001, more than a decade after the ICDP began. Before that three young people in Mananara-Nord town (Mananara-ville), who were proficient in French and English, had monopolized the guide trade. Several years prior to my arrival in Mananara-Nord, the three Betsimisaraka guides (a woman and a pair of brothers) had formed a guide association. Although it was often difficult for tourists to find one of the guides when they arrived in Mananara-Nord, they were required to have one to enter the national park. In the late 1990s, the three guides attempted to get regular employment with the ICDP. But the wages offered and the bosses' insistence that tip amounts be standardized annoyed the guides, and they ended up rejecting the offer. One of the guides complained that the biosphere project wanted to treat them like *guides sauvages* (savage guides). He meant that the wages, being much less than the fees and tips they might otherwise earn from tourists, implied that the biosphere project bosses considered them bumpkins, too rough and unschooled to please foreign tourists. Sylvestre, the conservation agent, had overheard one of the former guides tell me this story one day, and not having grasped the full context, he took offense at the term *guides sauvages.* He mistakenly thought that the project bosses had referred to the conservation agents, including Sylvestre, as "savage guides." (Of course, it was the former guide who implied that he considered conservation agents "savage" compared to himself, his brother, and the third guide in town who used to accompany all the tourists into the national park.)

Before the crew of conservation agents began to guide tourists through the park, they had occasionally been called upon by their bosses to accompany and carry the packs of scientists who were authorized to conduct research in the national park. From the guides' perspective, this job duty, like tourist guidance, was desirable because it exposed them to scientific practice. Many of the men aspired to learn scientific terms and methods as a means to climb the ranks of the conservation bureaucracy, just as they enjoyed interactions with tourists (but also felt uncomfortable) because they had the chance to practice and improve French-speaking skills and to show off their knowledge of faunal behavior and medicinal plants. They also might receive tips and other gifts.

Because of its access to the rain forest, the village of Varary possessed a passageway house (*gîte*) built by conservation and development agents. The *gîte* was supposed to be used by both ICDP employees and tourists. It was a simple plank and thatch room with a foam bed, a mosquito net, a wooden chair and table, and a detached outhouse. For tourists, it was disagreeable. The bed was full of mites, the pillowcases smudged black from the coconut oil Betsimisaraka women use in their braided hair. After a few days of disuse, cobwebs would line the thick reed walls, and cockroach droppings would pepper the floorboards and table.

I stayed in the *gîte* while in Varary for a couple of weeks each month, and I only twice had to vacate it to allow a tourist to stay. The reason tourism had slackened, according to Sylvestre and other residents, was because the ICDP bosses redirected tourists to a different entry point on the opposite side of the park on the coast. Although this was a longer trek for tourists, it was more scenic. The bosses said to the conservation crew that the footpath leading from Varary into the rain forest exposed too much unsightly deforestation, suggesting that over a decade, between 1990 and 2000, the landscape around Varary had changed dramatically.

Tourists who stayed in Varary usually spent one night so as to be able to begin their trek in the cool early morning. They brought large backpacks, digital and video cameras, bug spray, and heavy hiking boots. Sylvestre said that they tended to make their packs too heavy, they wore the wrong shoes, and they could not tolerate the heat of the sun for very long without feeling light-headed or, worse, nauseous. Hiking boots were inappropriate footwear because the paths often softened into deep mud pits or traversed coursing rivers. (For *vazaha*, or foreigners, light plastic gel sandals known as *kiranily* were the only good option.) Intent on enjoying the spectacle of nature, tourists were often ill-prepared to walk in it. Sylvestre recalled a time when he, acting as guide, led two tourists over a portion of the footpath that dissolved into deep sludge, over which were flung two long, rotting planks that served as a *passerelle*. The tourists stood still, helpless, and began to slowly sink into the sucking mud under the weight of their packs. Sylvestre turned around, hauled them onto the sunken planks, and piggybacked each one to firm ground.

On an outing in the biosphere reserve with Raleva, another conservation agent of the ICDP, I thought I was going to have to be carried or else left in a village to recuperate. It was during my first field trip into the reserve. I was accompanied by Raleva, a conservation agent in charge of disseminating information to peasants about a new community conservation initiative. Born of a Betsimisaraka mother and a Tsimihety father to a life of farming, Raleva, thirty-nine, grew up as a cultivator in a village close to Mananara-ville. Serious about his education, he attended junior high school in Mananara-ville and later obtained an electrician license. Raleva spoke excellent French and fluent Merina ("official Malagasy").

His assignment with the ICDP involved long, arduous treks to the remotest hamlets of the biosphere reserve. The ability to walk long and far was essential for ICDP field agents. In early February 2001, he was heading to the village of Savarandrano, about 30 kilometers from the end of the vehicular road in Sandrakatsy, where he would help residents create a community conservation association (McConnell and Sweeney 2005). The Contractual Forest Management agency (Gestion Contractualisée des Forêts, GCF) conceded management of state-controlled forests to village associations, whose members were given permission to harvest a certain amount of timber, plant fibers, and protein from the forest annually as long as they abided by sustainable management practices and laws protecting endangered species. Raleva was in charge of explaining GCF rules to villagers and drawing up contracts between village associations and the state.

As we slogged through sucking mud, goose-stepped heel-to-toe through narrow, hard crevices of dirt, and climbed the slick root-woven trails, at some point my knees seized up in revolt. For the remainder of our four-day trip, I hobbled along, wincing. Most of the conservation agents beyond the age of thirty-five, including Raleva, complained that they were too old for this kind of work, especially long walks on mountainous trails. Yet, compared to me, Raleva was fleet-footed and powerful. My presence with him on the mountain trails attracted a lot of interest. *Vazaha* were a rare apparition there. The sight of me prompted a favorite topic of conversation in the region—the incompetent walking of *vazaha*. *Vazaha* "don't know how to walk" (*tsy mahay mandeha*), people would say with amusement as they watched foreign tourists and researchers like me, winded and limping from a long trek on the tortuous footpaths. The act of walking in rural places informs people's knowledge of the land, forging intimacy with place, as well as their assessment of a person's value as a competent, able-bodied land-worker. Walking the land of one's ancestors establishes rootedness to the land and its features, which are animated by dead ancestors and replete with their memories. Being unable to walk proficiently exposes one's estrangement from place. It reflects a lack of belonging, an unvalued otherness.

I have introduced Raleva and Sylvestre to highlight the critical role they play in enabling contemporary tourists and researchers to navigate the forest and surrounding terrain, in mediating the encounters of insiders and outsiders, and in guiding visitors' perception of the rain forest, Malagasy society, and the landscape. Malagasy people's "embodied experience of pedestrian movement" contrasts with the historical experience of foreigners in Madagascar, who were neither expected nor really given leave to walk long distances on the open paths (Ingold 2004:322). Modalities of transport and perception inherent in the social division of precolonial Madagascar forged values that continue to be expressed through anatomical-spatial representations. These entail the association of social climbing with optical knowledge, and job stagnation with ambulatory knowledge, while in the village able-bodied mobility confers higher social standing than ambulatory impairment, and mobility is a trait worth protecting in the afterlife. These semiotics of social relations inform the structure and work-content of neoliberal conservation.

Today, the work of Malagasy porters and guides still enables outsiders to collect data about rain forest species, as well as the clandestine activities of animal poachers and peasant horticulturalists. Local Malagasy guides with intimate knowledge of the forest trails also assisted photographer Toby Smith in his undercover mission to substantiate rumors of the rosewood logging going on now in protected areas of the east coast.[1] Malagasy underlings have sought to elevate their standing in colonial and postcolonial bureaucracies in part by learning how to "see" species and landscapes, and to articulate this visual knowledge in the way outsiders do.

The worker stands on a higher plane than the capitalist from the outset, since the latter has his roots in the process of alienation and finds absolute satisfaction in it whereas right from the start the worker is a victim who confronts it as a rebel and experiences it as a process of enslavement.

—KARL MARX, *Capital*

La Grande Isle is a vast and rather poor country with a sparse population. Without work, without order, without a workforce, this country would remain in or return to its previous state before we took it in charge . . . an immense land, ravaged and sterile, on which indolent and scrawny populations live, decimated time and again by famines, shaking off their indolence only to cut each other's throats or pillage each other.

—HENRI PERRIER DE LA BÂTHIE, "Le salariat indigène à Madagascar"

3 Land and Languor

On What Makes Good Work

Shortly after Madagascar was annexed to the French Empire in 1896, the colonial administration began to make a concerted, large-scale effort to bring the forest of eastern Madagascar into capitalist production. Official foresters made reconnaissance missions to distant outposts such as Mananara-Nord to survey regional resources and to assess how best to exploit and transport these resources to ports. Extracting the rain forest's natural wealth would necessitate the development of a wage labor force. Betsimisaraka peasants would have to become dependent on wages rather than earning a living as independent cultivators. A constant problem for private entrepreneurs and colonial officials throughout Madagascar was the refusal of Malagasy subjects to offer up their labor. Betsimisaraka and Tsimihety horticulturalists and fishermen fled into the deep forest to escape taxation and coercion into industrial work sites. Colonial officials did manage to round people up, however, and by the mid-1920s the state instituted a forced labor regime called the Service de la Main-d'Oeuvre des Travaux d'Interêt Général (SMOTIG) in which conscripts ("pioneers") would serve the state for two years, and often private industrialists would "borrow" the pioneers from public work sites for their own logging, plantation, and mining operations, as they could rarely muster enough voluntary hands (Sodikoff 2005a).

Peasants' reluctance to take on wage work in cash crop plantations and timber concessions fed into racist assessments of Malagasy labor by *vazaha* bosses, including French colonial administrators, entrepreneurs of diverse ethnicities and nationalities, Chinese and Indo-Pakistani merchants, and Merina functionaries. The French were especially critical of the "phlegmatic" nature of the "east African" coastal populations of

Madagascar. Among these, Betsimisaraka people were considered congenial but lazy and governed by superstition. The Malagasy practice of *tavy* substantiated what was deemed an intrinsic laziness that exasperated the *colons*. The Merina and European idea of Betsimisaraka laziness made it difficult for the French and their collaborators to imagine that Betsimisaraka people possessed any kind of work ethic. Yet while Betsimisaraka men usually either resisted or entered into capitalist work sites reluctantly, under certain conditions they did so willingly.

This chapter focuses on the politics of value about work itself in colonial Madagascar, a politics inherent in capitalist workplaces throughout Mananara-Nord. Rather than focus on instances of Malagasy resistance to colonial authority, a dynamic that plays out in contemporary conservation efforts, I examine the acts of acquiescence and moments of compromise between Betsimisaraka workers and their "outsider" employers (those whose natal territories lay beyond their place of residence). The latter have sought Betsimisaraka labor for piecework and wage work. Determinations of the right way to work and the right way to exploit land have transformed the landscape of Mananara-Nord from a district in the thick of the contiguous northeastern forest of at least 120,000 hectares in 1900[1] to one containing less than a fifth of that at the core of a biosphere reserve nearly a century later.

LABOR RELATIONS IN A MANANARA-NORD HOUSEHOLD

While the idea of "the lazy native" was widespread in tropical European colonies (Redfield 2000:217), it has also had currency amongst Malagasy groups. The "tribal" classification scheme of the French borrowed from ideas that already existed among Merina elites (Decary 1958; Valenksy 1995). Although the French had plenty of negative things to say about the lethargy of the highland populations as well (namely the Merina and Betsileo ethnic groups), they also expressed admiration for these resourceful "asiatics," who built earthen terraces on the hill slopes for the production of paddy rice. Today, Merina settlers in Mananara-Nord point out that unlike the *mazoto* (industrious) peasants of the highlands, Betsimisaraka peasants eschew the intensive labor of building soil-protective terraces and irrigation canals on their plots, preferring the "easier" route of razing the forest when the fertility of fallow land diminishes.

In the town of Mananara-Nord, often called Mananara-ville to distinguish it from the prefecture name, non-Betsimisaraka residents constantly talked about the quality of Betsimisaraka labor. "Betsimisaraka don't know how to work" (*tsy mahay miasa*) was a habitual complaint of Merina settlers (Feeley-Harnik 1984:3). They would often recite Betsimisaraka workers' faults: Betsimisaraka came late to the job, quit without notice, absented themselves for days after being paid, took no pride in their tasks, were lazy, vulgar, and learned slowly. Betsimisaraka also practiced *tavy,* or *jinja,* as they call it in the northeast. I had heard a somewhat watered-down version of these same complaints from North American and European expatriates who worked in the conservation and development sector as well, while Malagasy employees of ICDPs in Mananara-Nord and other east coast localities often complained bitterly of their treatment by bosses, especially those of Merina ethnicity.

In October 2000, I settled into the home of a Merina family, the maternal uncle, aunt, and niece of my friend Haja. Living with Lala, a seamstress, Mandresy, a lapidarist, and their daughter, Fanja, gave me an intimate glimpse of the dynamics of Merina-Betsimisaraka relations. During the course of my research, I spent about ten days of every month with this Merina family in Mananara-ville. The rest of the time was spent in the village of Varary, located near the western access point of the biosphere reserve's national park, or in other villages as I followed the movements of conservation agents.

During the first six months of my stay in Madagascar, Mandresy employed two Betsimisaraka men to manufacture crystal balls and other crystal ornaments for quartz dealers in town. Lala managed the duties of the maid, a young Betsimisaraka woman. The wages demanded by domestic workers in Mananara-Nord (approximately 150,000 FMG, or US$25 per month) were three times as much as those in Antananarivo.

The family was very active in the Protestant Church, FJKM.[2] Their devotion had begun after Mandresy experienced a revelation, gave up drinking and cigarettes, and took up the call to evangelize several years earlier. During my first visit to their home in 1999, Haja and I remarked on their habits early one morning, whispering, as we listened from the bedroom to Lala and Mandresy holding their daily "voluntary" Bible study with their two crystal ball workers a half hour before the start of their regular workday at 7:30 AM. Together, Lala and Mandresy would take turns reading a passage aloud with the same patient, pedagogical tone they used when discussing the Bible with their eight-year-old, Fanja. Then Mandresy would explain the text and its relevance to everyday life. Mandresy and Lala felt that Bible study and conversion to Protestantism would improve the men. They assured the workers that faith and prayer would bring them courage and prosperity. To us they explained that the Bible study sessions were meant to "civilize" Betsimisaraka people, to discipline them into being more reliable and diligent help.

When I returned to the home of Mandresy and Lala the next year to start my extended research, their household finances had taken a turn for the worse. They struggled to make ends meet. Gone were the two crystal workers who had listened to Bible stories. Without them, Mandresy had to handle all the quartz processing single-handedly. Lala, in addition to her sewing, had to clean, shop, launder, and prepare meals. Cooking alone was very time-consuming in a household that lacked running water and kerosene (they used a charcoal stove). Mandresy told me that first one crystal worker and then the other had "gone crazy" (*lasa adaladala*). Mandresy and Lala were sad about it and explained that it was because of the wavering faith of the workers, their doubting. Their inability to fully accept Jesus Christ had made them vulnerable to evil spirits (*jiny*), so they went crazy. Like numerous Malagasy Christians I met, both Merina and Betsimisaraka, Mandresy and Lala believed that sincere and repetitive prayer, Bible reading, churchgoing, and good works such as evangelization kept evil at bay, made one receptive to miracles, reversed bad fortune, and would ultimately fulfill wishes.

Lala and Mandresy disparaged Betsimisaraka labor but depended on it. Mandresy needed men to grind and polish raw quartz crystal, fashioning it into balls or faceted ornaments. Lala needed a sewing assistant as well as a maid to clean and cook so that she could sew. The family could only sporadically afford to hire help. Not too long after my arrival, the family was able to acquire new workers. Lala, otherwise warm and generous, acted the strict disciplinarian with Betsimisaraka workers in the house. Mandresy was affable with his workers but not collegial, wishing to maintain authority as a boss and moral guide. The couple believed that acting overly familiar with Betsimisaraka workers would result in their taking liberties. Lala thought severity maintained a necessary structure for keeping workers diligent and honest. This structure was implicitly forged and upheld by "putting down" (*manambany,* "to make low") the workers. During weekday lunches in the house, for example, the Betsimisaraka workers would sit in the back room, separated from the family (including me), who sat in the front room. The chipped ceramic and scuffed enamel plates, the cups with broken handles, and the rough metal tablespoons went to the workers. We used the good plates, stored safely in a wooden box, the finer china cups, cloth napkins, glasses, and the shiny silverware. The workers' plates were heaped high and wide with rice. Each would receive his or her individual bowl of *loaka* (the Merina word for the food that accompanies the rice, giving it flavor; called *rô* in Betsimisaraka). The workers found this preferable to having a common bowl out of which they would help themselves because, they explained, this made it hard to determine if everyone got equal portions. A parsimonious amount of the tougher cuts of meat, pieces with ligaments, cartilage, and chunks of bone, together with generous ladles of broth, filled the workers' bowls. We got the choicer cuts and more copious vegetables. "They like it that way," Lala told me. She was not wrong exactly. My initial attempts to even the score as I set the tables or doled out the *loaka* seemed to upset the natural order of things, making the workers uneasy and prompting Lala to correct me.

Lala constantly scolded her sewing assistant and the maid when they were newly hired. Both were young Betsimisaraka women from neighboring villages raised as cultivators (*mpamboly*). A visiting doctor friend of Mandresy and Lala, a Merina man, one day bluntly told Lala that she put down the workers too harshly (*manambany be,* "to really make low"). He figured this explained the couple's inability to retain their employees for very long. Mandresy agreed that Lala "needed to fix [her] character" with respect to the workers. The doctor argued that Betsimisaraka must be treated "softly" (*mora mora*), if one wanted good work out of them. Lala was surprised and chagrined at the assessment of her behavior because even though she believed in disciplining workers, she never intended to overdo it. She immediately then changed her manner. Using friendliness rather than anger, she succeeded in retaining all four of the employees for a record length of time, and the workers seemed relatively content at their jobs.

Mandresy had a reputation in town for being the best crystal processor. A mechanic by trade, Mandresy built his own grinding and polishing machines, which stood in the backyard and kitchen area and rumbled from dawn until dusk. Mandresy attributed his good reputation in town to the comparatively shoddy work of the few Betsimisaraka crystal processors in Mananara: they did not know how to grind as smooth a ball or achieve as high a luster; they were slow; they were not serious (they drank on the job); and they did not know how to cut the quartz to show off its special features, such as inclusions of fuzzy minerals, hairlike rutile, or trapped beads of water. Although he found his own Betsimisaraka workers to be tolerably good, Mandresy often dreamed of bringing in Merina workers from Antananarivo, the capital, to help him. "Then it would go really well here. Merina know how to work. Merina really know how to work rock," he would say.

On the other hand, he admitted, Betsimisaraka knew how to find crystal, called *vato mahita* ("rock for seeing," or "see-through rock").[3] Peasants in the countryside were expert at finding places where quartz crystal lay embedded in the earth. When covered with dense clay, raw quartz is indistinguishable from opaque rock to the untrained eye. Rural peasants would occasionally hunt for pieces of quartz to sell to buyers for extra cash. They would walk the terrain and poke the ground with a *baromeny* (a meter-long metal rod), listening for the particular "ting" quartz makes when struck. Although the Mananara-Nord region possessed high-quality quartz, including optical quality for lens manufacture, its quantity was diminishing. Because quartz was harder to find, Mandresy explained, fewer peasants devoted their time to crystal hunting in 2000 than had done so in the years before. Occasionally, peasants from distant villages would arrive at our house to sell crystal to Mandresy directly. A rigorously honest and relatively naïve man, frequently a victim of others' dishonesty, Mandresy was still a bargain hunter. He tried to get the most advantageous prices for high-quality or unusual pieces of quartz, justifying a good profit by saying that Betsimisaraka from *ambanivolo* had little use for cash and only wasted it recklessly when they had more.

With the easing up of the atmosphere, the relaxing of expectations, and the softening of Lala's tone after the doctor's criticism, Mandresy's workers, Honoré and Justin, worked consistently and grew used to—even amused by—the criticisms of Mandresy's few but regular Merina clients who brought their crystal to the house for processing. The livelihoods of Mandresy's workers depended on the inflow of quartz to the household, since they earned by the piece. The domestic workers, in contrast, received monthly wages. Bôtine, the sewing assistant, shortly after being hired grew bolder in asking for gifts and favors, and she frequently failed to show up on days she said she would. Berthe, the maid, also failed to show now and then without warning and started spending an inordinate amount of time at the beach after lunch, in Lala's view (and I agreed). Yet Bôtine had become adept at sewing on buttons, basting hemlines, and cutting cloth to pattern, and Berthe continued to be diligent in scouring the wood floor with the coconut husk underfoot every morning until the surface gleamed. Lala disapproved of Bôtine's impoliteness (*tsy mahalala fomba izy,* "she doesn't know good manners") and of Berthe's increasing laziness. But she was afraid to reprimand them because they might leave, and the household depended on their help. A balance of sorts was struck between the employers and employees in our household. Rather than putting workers down, Mandresy and Lala were compelled to lower the bar regarding punctuality, daily reliability, time efficiency, and their notions of appropriate behavior. The employees, in turn, by reacting negatively to an atmosphere of scolding and mutual distrust, and by interacting more familiarly with their employers and with each other, improved the "ambiance" of the workplace/home.

Mandresy and Lala's household reflect the ethnicized structure of labor rooted in Merina imperialism. *Mazoto* (industrious, tireless, ready to work) and *kamo* (lazy, sluggish, unwilling to work) are terms that present-day Merina on the east coast use to compare their own work habits with those of Betsimisaraka (Feeley-Harnik 1984).[4] Betsimisaraka people's "laziness" has been attributed variably to either their disdain for or their essential inability to work, which frustrates employers and validates their sense of superiority. Mandresy's opinions about Betsimisaraka skills in crystal hunting and processing demonstrate one small way in which ideas about landscape and labor get expressed in one breath (see Harper 2002).

When Merina deem Betsimisaraka lazy, they echo French colonial administrators who sought to instill in Betsimisaraka a sense of anxiety over wasting productive hours and potential forest commodities. The friction within the household among the workers over wage labor versus piecework echoed disagreements over the proper way to value labor a century earlier. Wage labor and later forced labor were considered means to lessen peasants' reliance on slash-and-burn horticulture. The state attempted to concentrate Betsimisaraka peasants into larger villages, rather than dispersed settlements, to facilitate tax collection and labor recruitment, a policy that encouraged peasants to establish villages that were hidden deep in the woods. The ban on *tavy* was an additional attempt to squeeze peasants onto cash crop plantations and timber concessions, where their coming and going was more easily monitored by supervisors.

THE GEOGRAPHY OF LANGUOR

Written assessments of Betsimisaraka labor in Mananara-Nord date back to 1900, the year in which M. Jeannelle, the general guard of forests, reports on his foray into the "great mountainous forest." Colonial French observed Betsimisaraka peasants as they established and tended their rice plots, and later as they were forced to assist in the development, or valorization (*mise en valeur*), of the colony. Although many administrators grasped that Betsimisaraka did not like working for *vazaha*, they attributed this dislike to a deficient or nonexistent work ethic. The French described a Betsimisaraka *force de l'habitude* (force of habit) in terms of *la paresse* (laziness), *l'indolence* (indolence), *l'apathie* (apathy), and *l'oisiveté* (idleness), tied to a general *insouciance* about the future.

French colonial reports decry the "force of inertia" affecting Betsimisaraka as the main obstacle to the region's economic development. The conception of this lazy "race de primitifs"[5] infused how colonial officials saw Betsimisaraka territory: as a lush and fertile forest victimized by a slothful slash-and-burn agricultural system. Annual forest clearance and rice sowing were considered destructive habits "rooted in" (*enraciné*) Betsimisaraka and Tsimihety peasants.[6] Colonial officials and entrepreneurs sought manpower to extract and process timber, fibers, ornamental plants and crops, animals, and gems and other minerals. They had to penetrate the "force of inertia" if they wanted to bring Madagascar's human and nonhuman nature into fruitful production.

In 1900, Jeannelle was sent to Mananara-Nord on a reconnaissance mission. He subsequently submitted a report to Governor-General Gallieni containing twenty-one pages worth of topographical and ecological information, intermixed with a few ethnographic passages about the district's Betsimisaraka inhabitants.[7] Jeannelle's main goal was to map the territory's economic potential, so his report describes regional floral species, the kinds of cash crops amenable to these soils, and the water networks usable for the transport of timber and plants from the inland mountains to shipping ports. His ethnographic observations are limited to Betsimisaraka practices involving tree extraction or conservation. He remarks, for example, on the beautiful, rare tree specimens that Betsimisaraka people fell, "without worry over their location and the difficulty of transport," to build canoes that can contain up to fifteen persons. He notices isolated thickets of forest standing in proximity to the grand forest that "owe their conservation to different causes." These woods may either be "populated by spirits" or serve as graveyards for Betsimisaraka.

Jeannelle's assessments of Betsimisaraka labor reflect the scientific knowledge of a geographical determinist (Redfield 2000). The humid, luxuriant environment cultivated a primitive, torpid people. He does not dwell on adaptation or on the correspondences between Betsimisaraka and their habitat, but instead dramatizes an antagonistic relationship between the wilderness and the natives. His dour view may owe to his physical exertions over the region's "primitive" routes, forested mountains, sandy coasts, fierce waterways, and "nocuous miasmas." Finding the terrain obstructive and excessive, he recalls squeezing through the forest where the paths have been eroded into 2-meter-deep corridors "hardly as wide as the human foot." The eastern-facing mountain slopes "rise insensitively" to an altitude of around 1,200 to 1,300 meters, and the slope is "infinitely broken down" by secondary mountain chains. Rivers are cut by an "infinity of rapids" that pose "grave inconveniences" to transport by water. The poor condition of forest trails, which are covered with slick clay and embedded with crystalline schist and tree roots resembling "stairways," and the treacherousness of the rivers complicate the task of timber removal for *concessionaires*. He remarks on how the abundance of rain on impermeable alluvial deposits makes for a "malarial region *par excellence.*"

French administrators conflated their ideas about Betsimisaraka land and labor at the sites of charred swaths of land at the edges of forest. Maligning the "barbarous customs" of peasants, Jeannelle portrays burnt land in language that evokes human carnage: a land "strewn with incompletely carbonized wood; all over lie parts of trees at one's feet, half burnt, exposing their blackened trunks." The destructiveness of *tavy* is bound to the Betsimisaraka proclivity for "killing time" (*tuer le temps*). The view of Betsimisaraka as a people overcome by laziness afforded Europeans an easy explanation for their local labor problem. In a telling turn of phrase, one British historian captured the view of French administrators bent on spurring the "deadening enervation" of the native; such torpor seemed to "creep over Malagasian affairs with the dreadful inexorability of a tropical forest encroaching on cleared land" (Roberts 1929:411). He continues: "The *indigènes* prefer to waste time than to work without care to improve their miserable existence. So they ravage the forest with slash-and-burn agriculture, a method requiring minimal labor time."

But because Betsimisaraka peasants were best adapted to the eastern forest, colonial administrators found them to be essential human resources. In Mananara-Nord, Betsimisaraka possessed boating expertise to convey materials by canoe to the coast and could "thread their way into [*faufiler*] the massifs with surprising agility." The behaviors of Betsimisaraka that colonial administrators would attribute to laziness—reflected in the apparent lack of effort with which Betsimisaraka existed in the forest—would prove indispensable to the process of land valorization.

In the European view, capitalist wage work offered a remedy for Betsimisaraka "laziness." Colonialists believed that a change in the mode of production would improve Malagasy people's moral worth and would simultaneously enable the forest to recuperate from the assaults of *tavy*. The capitalist work ethic as it manifested in rain forest regions was implicitly conservationist, notwithstanding the fact that colonial production decimated one-third of the country's 12 million hectares of "exploitable forest" in the first half of the twentieth century (Boiteau 1958:227–232). French officials downplayed or crossed out references in reports to the impact of capitalist enterprises on the eastern forest.

KILLING TIME AND TIMBER

From 1906 onward, annual reports from Mananara-Nord complained regularly about the lack of labor for the agricultural and forest concessions.[8] The "lack" referred implicitly to Betsimisaraka men's will to work rather than to the number of able bodies present. Year in and year out, Mananara's administrators assessed each of the economic categories they sought to develop—animal husbandry, agriculture, commerce, forests, and labor—and grieved about the lack of a diligent labor force as the key obstacle to Mananara's enrichment. M. Spas, the district head of 1910, recounts:

> In dreading the arrival of any European functionary, [Betsimisaraka] always produce . . . the disagreeable impression of grappling with a force of inertia that makes one despair: rare are the real workers, all produce but the strict minimum, and none tries to elevate himself above the others and make for himself, through work, a less miserable existence than this that is currently theirs. It's not even possible to appeal to their self-respect, and attempts towards this goal also appear as lost efforts. . . . It's this indolent and apathetic character that makes this Betsimisaraka race so rebellious toward all progress and so difficult, despite its gentleness, to manage. The Betsimisaraka is for the moment incapable of the least effort.[9]

In his denigration of Betsimisaraka labor, he does not imagine that "self-respect," not to mention a "less miserable existence," might in fact depend on a different moral code, a refusal to "elevate oneself above the others." In a 1912 report, Stefani writes of the difficulties *colons* faced in procuring cheap, reliable, local labor: "The natives show themselves to be very demanding at times. Very lazy, they don't find any need to give themselves over to regular work; except for the cultivation of rice for their personal needs, it is nearly impossible to pull them away from their apathy."[10] The "very demanding" attitude of Betsimisaraka workers, regarding the amount of wages for which they consent to work, exposes the limits of colonial efforts at persuasion as much as it demonstrates Betsimisaraka's valuation of their own labor.

The strength of the belief in native laziness dulled colonial administrators' powers of perception. They saw nearly all Betsimisaraka labors as demonstrations of minimal effort, which in administrators' view befitted tropical forest dwellers. The indulgent forest, containing "all the necessary materials for the rapid and painless construction of a house,"[11] enables a lifestyle suited to the "gentle" and "lazy" Betsimisaraka character.

In reports, the Betsimisaraka's proclivity for "killing time" and ravaging the forest were perpetual frustrations for French administrators:

> letting the days go by in perfect indolence, in the most complete insouciance of tomorrow and the future, wasting their time in reveries, in interminable *kabary* [customary speeches], in visits to their families, friends, etc. The local labor is, it follows, frankly bad: unintelligent, demanding regarding salaries, irregular, barely productive . . . working not in view of some goal specified by the employer but to "kill time," to reach the end of the day or month.[12]

Killing time equaled killing trees. *Tavy* epitomized Betsimisaraka's mismanagement of time and disorderliness in their work habits.[13] Stefani writes:

> The Mananara native cannot understand and does not want to understand that work must be ordered, continual, and that one's interests motivate the doing of certain works in advance. I have not yet been able to make enter into their firmly closed heads that the preparation of a plot or paddy, the cutting, the clearance, the water passages, can and must be effected in advance, in a manner that would have nearly no work at the moment of sowing or planting—instead of this, the Betsimisaraka, like the Tsimihety, always irresolute, incapable of stopping to consider the evening before what they will do the next day, decide all of a sudden to plant rice or something else: quick, the terrain is burned, enclosed, cultivated, in one fell swoop, without rest, because hunger makes itself felt. It's the temporary effort of the "lazy who set to it" and who then wait patiently for harvest.[14]

For the French, whose sense of pace outstripped the apparent "natural" pace of Mananara-Nord, swidden agriculture reflected the stasis of Malagasy ancestral custom (*fombandrazaña*). Doing *jinja* was perceived as historically unevolved yet radically shaping the island's natural history.

Colonial administrators' and concessionaires' anxiety over the land being laid to waste was rooted in a clock- and profit-disciplined temporality, precisely what Malagasy seemed to lack (Cooper 1992). Tasks of necessity oriented Betsimisaraka time and commingled "social intercourse and labor" (Thompson 1967:60). The French sought to reform Betsimisaraka time-space through a combination of a stricter enforcement of rules against forest burning, regular wage work, and forced labor.

THE FORCE OF HABIT AND THE HABIT OF FORCE

Males of the Mananara-Nord region were introduced to compulsory labor policy in 1901, when they were needed for projects much farther south, at the work sites of the Antananarivo–to–East Coast (Tananarive–Côte Est, or TCE) railway construction.[15] In June 1902, Mananara-Nord's district head, E. Lagriffoul, was obliged by Governor-General Gallieni to find laborers from Mananara to work for a period of two months at railway sites moving westward from the eastern port of "Anivorano" [*sic*]. It was an awkward situation for Lagriffoul because labor was already scarce for the concessionaires of Mananara, and Betsimisaraka residents were not receptive to working for *vazaha* (Cole 2001). Lagriffoul wrote that he was in the process of initially sending southward a hundred "voluntary workers" on a steamship, the *Ville de Pernambuco.*

Lagriffoul's use of the term "voluntary" was deceptive. It suggested the administration's desire to distance itself from the imposed slavery of Merina rule prior to France's colonization. Joseph Mullens, a British foreign secretary of the London Missionary Society, described the *fanompoana* (obligatory labor for the sovereign) system in 1876, before France's colonization of Madagascar: "It bears heavily on the skilful, it is unequal in its demands, it represses progress by taking away any stimulus to self-improvement, or to individual enterprise. It keeps society on a dead level, and fosters indolence and indifference. It will only be cured by the fair distribution of required services within all grades of society, and by the commutation of service through a fixed money payment" (Mullens 1876:195). Yet Gallieni was routinely exacting from Malagasy subjects their *prestation* labor for public works, *prestation* being a form of forced labor that recalled the *fanompoana* instituted under the Merina monarchy (Campbell 1988; Feeley-Harnik 1991). In Lagriffoul's 1902 letters to Gallieni, he grapples both with the command to find and send workers south and with trying to manage the labor and food needs of Mananara.

He intimates the stresses of his own labor as manager, rendering himself a benevolent patriarch, as he berates the nature of Betsimisaraka labor:

> The departure of workers is a big effort for the Province. Betsimisaraka, in effect, are timid, and consent to work, but near their home. Voyaging far from their country causes them some apprehension which I have sought to dispel in numerous *kabary* [formal speeches]. . . . Also I would be grateful if you could recommend to the directors of the railway works to treat the Betsimisaraka according to their character, that is to say, rather with gentleness. They are capable of providing a good labor if they are coaxed, and one must be with them as with children. Their example will be otherwise salutary in the Province, they will inspire confidence in others, and in other times, when the rice crop is finished, I will be able to send workers regularly.[16]

How in fact did Lagriffoul satisfy the demands of the governor to provide labor for public works in the midst of the rice harvest, when Betsimisaraka peasants devote long hours to cutting each panicle of rice off its stalk, one by one, and hauling heavy bundles of rice to their silos? How did he manage to find a hundred willing laborers in the face of the Betsimisaraka people's strong reluctance to work for *vazaha*, or to work outside their natal region? Nearly four months later, Lagriffoul wrote a letter to the governor-general in which he tried to dispel a rumor that he had "forced eighty workers" to embark the *Pernambuco* at Maroantsetra.[17]

If at the outset administrators gingerly sidestepped mentioning force as a matter of policy, it quickly became the only conceivable means of combating the "force of inertia" permeating the region. One way that the government attempted to set the Malagasy to work was through the imposition of taxes, with "voluntary labor" described as a means of payment. In 1911, Stefani, the head administrator of Mananara-Nord, recommended the institution of forced labor to address the problem of laziness, thereby imposing both capitalist time-discipline and disciplined forms of land use onto Betsimisaraka subjects. His rationale makes explicit the goal of rescinding Betsimisaraka people's freedom. He writes:

> The day where the native would be forced to work every day, like the European, and that this obligation would become a habit, what profits would colonization gain. . . . In forcing the natives to work 8 or 10 days per year, we would easily see indispensable communication routes: it does not seem

> possible to recruit labor other than by banishing, with a decree, the idleness amidst the natives, still primitive and savage people, for whom independence is not a good thing.[18]

A subtle choice of phrase, "idleness *amidst* the natives," rather than "*of* the natives," may hint at a shift in colonial administrators' outlook. If the quality of idleness infused the milieu of Betsimisaraka people, like a miasma or virus, but not necessarily their essence, it might be rooted out or purged through a regime of forced labor. In light of their failure to discipline Betsimisaraka peasants to the routine of regular concessionary and plantation labor, and to sedentary, intensive agriculture, French administrators treated forced labor as a kind of apprenticeship: first impose the discipline, and later, through voluntary labor, offer remuneration.

The extent to which forced labor served its disciplinary purpose is reflected in subsequent administrative reports in which administrators complain about the persistence of *tavy* and laziness.[19] Gillian Feeley-Harnik writes that in 1908, Gallieni assumed "that the best spur to native indolence was the imposition of taxes"; by 1930, that "solution was thought to have been pushed to its limit" (1991:127). Between World Wars I and II, the state increasingly relied on forced labor through the military-like institution of the Service de la Main-d'Oeuvre de Travaux d'Interêt Général (SMOTIG) (Covell 1987), or through *Travaux Publics* projects.

In Mananara-Nord, peasants' oral histories of colonial forced labor not only reveal the extent to which the experience of forced labor is negatively remembered in the present, but also invert colonial conceptions of laziness, insouciance, rational land use, and time-discipline. In 2001, residents of the inland village of Varary—originally named Añalalava, "long forest," over 150 years ago—remembered, for example, their obligation to carry *vazaha* over the rough terrain of Mananara on *filanjana*. While colonial officials saw Betsimisaraka as insouciant about tomorrow, Betsimisaraka remembered *vazaha* insouciance about the humanity and welfare of Betsimisaraka subjects.

Contemporary Betsimisaraka also remember charges of rule breaking, a crime of which they were regularly accused, particularly with regard to the ban against *tavy*. By 1932, Malagasy people convicted of burning land without authorization were penalized with six months in prison and a fee of 200 francs.[20] The president of the *fokontany* of Varary, a *fokontany* representing the administrative unit of a hamlet,[21] recalled in January 2001 his father being imprisoned for one month for burning land without authorization. He claimed that the inmates' work involved cleaning the outhouses of everyone (including "*vazaha,* Chinese, Malagasy, Betsimisaraka"): "They had to dump [the contents] in the sea with their hands. . . . The guardian of the prison led prisoners at night to the WCs [*kabone*] around town. *Vazaha* gave soap for the work. They washed in the river near the sea in Mananara-town" (Field notes 1/13/2001). As with the *mpilanja,* recollections of contact with the excrement and urine of *vazaha* (and others) convey the historicity of forced labor, or the mood of social memory: an experience of ultimate debasement.

According to Betsimisaraka residents, inconsistencies in the administration's adherence to its own rules, in addition to concessionaires' unreliable payment of workers' wages, belied expressions of moral rectitude in administrative documents. Lema, the father of one of the Varary-based conservation agents, Jafa, recalls, for example, the way taxes were levied: "If one doesn't pay taxes, one is imprisoned and tied up. But if you were imprisoned during the time of *vazaha* [French rule], one doesn't pay taxes, because it is paid off in jail. But it was shocking, there were times in our day when at the same time one would be imprisoned and then have to go find money for taxes" (Field notes 3/10/2001).

While French officials distinguished forced labor from slavery, the former being in the "public interest" and a temporary obligation, Betsimisaraka denied this distinction. As I mentioned in the previous chapter, Rasonina, the *tangalamena* of Varary, compared forced labor to slavery under the Merina rule. Born in 1929, Rasonina recalled in February 2001 his own father's stories of peasants building the road between Mananara-Nord and Maroantsetra, as well as the hospital, airfield, and post office in the 1920s and 1930s:[22]

> With the *vazaha* there was slavery too. It came from Soanieranivongo towards Maroantsetra. You, you, you [he gestures toward imaginary persons] would all be put in a bag, one large bag. . . . For example, take me, I'd be your carrier. Two of us would carry you on our shoulders all the way to Maroantsetra. . . . Yes, you have to go all the way to Maroantsetra . . .

> The *vazaha* said that the Malagasy would suffer because there needs to be a post office and an airfield. So the law was made, and it was strict. It took young men, strong ones, twenty to thirty years old, and for six months they would work the land for the airfield. Oh oh oh oh oh!! Ah, they'd say, "please, those of you from Mananara, it's only for this one time, and then we'll be done with the airfield—and the road." . . . We made the road. This was hard labor too. How long were we there?
>
> So during this work on the road, we'd use something like a spade, a pick axe . . . what's the name of it? . . . An iron rod, a hammer for breaking the rock, a pick axe like the horns of a zebu, two to dig the soil. . . . The guard oversaw us. . . . "No resting," they'd say, . . . "it won't kill you." They didn't feed you; you had to go get your own food. Ah! You might be away for two days—it took a long time. If someone had a foolish wife here he'd starve to death. There was nothing to eat. The workers of the land [*mpiasa tany*], the peasants [*tantsaha*], with their spades, and pick axes, and rakes, and sledgehammers . . . would sometimes come across a river, a big river, that needed a bridge. We also had to build that. [The *vazaha*] would say, "make one." . . . The *vazaha* practiced slavery. That's not at all true that they stopped the suffering [of slavery under Merina rule]. It was clear the *vazaha* aimed to make a road and the airfield. Yes, yes, yes, they practiced slavery. Oh! They practiced slavery, if not exactly that. If, for example, slaves were working working working, slowly working, then those knives with the small handle may appear, that knife with a wooden handle . . . you'd pass out from dizziness and there they were! (Field notes 3/10/2001)

Rasonina's account conceptually merges the "time of Hôva" (*fahahôva,* Merina rule), which was a period of slavery, and the "time of *vazaha*" (*fahavazaha,* French rule), an era of forced labor. The reference to "being put in a bag" reiterates stories I heard about the capture and enslavement of Betsimisaraka by Merina. This memory is grafted onto that of carrying *vazaha* by *filanjana.* Both "times" involved Betsimisaraka relinquishing their labor at the threat of violence. Betsimisaraka villagers typically held *vazaha* in greater esteem than Hôva even though they were equivocal on the matter of whether or not *vazaha* "liked the Malagasy" (referring to Betsimisaraka, since Betsimisaraka do not consider Merina to be true Malagasy).

Rasonina's account invokes other affronts: colonialism turned Betsimisaraka against Betsimisaraka. At one point, he mentioned a man named Tsimanjo, an overseer (*surveillant*) who carried a "Malagasy-style stick" and physically abused Betsimisaraka workers. The memory of a Betsimisaraka person turning on his people was condemned as a deep betrayal. (Rasonina's voice grew louder, his eyes beseeched ours with incredulity as he mentioned Tsimanjo; he shook his head.) In my daily interactions with people in Mananara-Nord, no one ever talked about the significance of the ethnonym Betsimisaraka ("the many who do not separate"), but the disapproval expressed over these kind of anecdotes and experiences of betrayal conveyed a collective ethos otherwise glossed as *fihavanaña* ("familial relations"), a pan-Malagasy concept of treating in-group members, whether kin or nonkin, with loyalty, in solidarity, as family.

The *tangalamena* narrative also expresses a sense of time slipping away: "How long were we there?" he asks. Episodes of blacking out and waking to knife point stress the experience of time ill-spent. His account suggests the "contestation over time" occurring between colonial officials and Malagasy peasants, and it offers justification to resist the colonial notion of time-discipline (Cooper 1992:211). Compare Rasonina's comments about the Malagasy's sense of time to those of Europeans who learn bits of lore about Malagasy, such as this anonymous, presumably North American, commentator in the March 23, 1942, issue of *Time* magazine, who in an editorial contemplates whether or not the Vichy state would be forced to fight to maintain control over Madagascar, keeping it out of the hands of the Allies. The author begins the piece with the erroneous claim that the

> soft-eyed, soft-voiced people of Madagascar speak a liquid language in which there is no word for time. Under French administration, however, most of the Malagasy (as ethnologists call them) have had a few years' schooling. Some have even taken posts in the Colonial Government and learned the European world view. They know what time is and how fast it can run. Last week they wondered how Madagascar fitted into the Axis time scheme. (*Time* 1942:36)

While the Malagasy language does in fact have a word for time (*fitoana*), the assertion that France bequeathed onto its colonial subjects the concepts of time, of time running out, and the time of a world capitalist system that necessitates engagement in battle over space and the natural wealth it contains reiterates Western conceptions of the tropical periphery. This commentator (nationality unknown) interprets Malagasy people to exist literally beyond time (in a "world out of time"), unfamiliar with the abstraction of time itself and therefore off the conveyer belt of progressive history. That is, until colonization, a process of violently imposed modernization or contemporization.

PIECEWORK

Although the state rationalized its compulsory and forced labor demands in the name of the public good,[23] such colonial labor regimes were internationally criticized by the 1930s (Cooper 1996; International Labour Conference 1939).[24] At the same time, a popular insurgency triggered by the ban against *tavy* in Mananara-Nord also worried administrators.[25] While the more egregious forms of forced labor were phased out by the mid-1930s, the institution of compulsory labor persisted after Independence in 1960 and during the first Republic, when, Betsimisaraka residents recall, men had to submit to ten days (*folo andro*) of public works every year.

In spite of colonial administrators' claims to the contrary, evidence exists of Betsimisaraka men willingly seeking out piecework in Mananara-Nord in the early twentieth century. State-imposed taxes and penalties for tax evasion made cash a necessary commodity for peasants in Mananara-Nord, as everywhere else. While some Betsimisaraka and Tsimihety residents during the colonial era managed to escape into the forest to avoid taxes and forced labor, those (the majority) who chose not to abandon their villages needed to find cash. The pursuit of cash is part of contemporary life for Malagasy villagers. In Mananara-Nord in the early 2000s, they spent what little they have on clothing, cookware, medicine, staples (kerosene, matches, sugar, and salt), goods considered luxury items (boom boxes, foam mattresses, corrugated iron sheets for roofing, and bicycles), and rice during the lean months.

The colonial state began granting forest and agricultural concessions to French settlers and companies in Mananara-Nord in 1901. By 1903, a total of 121,690 hectares had been granted to seven European concessionaires for exploitation.[26] These concessionaires depended heavily on local labor to extract trees and to convey timber by river from inland to the coast, where it could be shipped south to the main port of Tamatave. In the early 1900s, the usual wages for local laborers were

60 to 70 centimes daily plus rice, which officials estimated at 20 to 30 francs per month, or 240 to 360 francs annually.[27] In comparison, the personnel at one concession in 1906 included a director who earned an annual pay of 10,000 francs, an assistant director at 6,000 francs, two Creole (presumably from Reunion island) clerks at 1,000 francs, and a mechanic at 3,000 francs.[28] Such low wages for Betsimisaraka prevented workers from being able to save and offered little incentive to forgo subsistence farming and foraging. Scrounging for money by selling odds and ends and offering one's labor for various tasks is called "seeking the leaves of weeds" (*mila ravin'ahitra*). "Weed leaves," like scratch, symbolize the paltriness of what is gained.

The majority of concessionaires in Mananara-Nord had difficulty obtaining labor. One exception was a M. Le Conte of Antanambe, who as early as 1901 had a constant flow of itinerant workers willing to fell large trees and transport them to his hangar, despite the fact that he paid the same "reasonable" wages as the other concessionaires. From the earliest years of the remote outpost's existence, district heads remarked on Le Conte's relative success in obtaining labor. In 1901, the head district administrator, Royet, deeming Le Conte "the most amiable and best of our *colons*," writes: "M. Le Conte bids, for example, for five thousand railway ties, he calls the habitual workers and says to each of them: 'I give you 1.2 francs per sleeper of these dimensions and of this wood, every Saturday I'll settle up with you that which you have brought me.' And in this way M. Le Conte can meet all the orders that come from Tamatave."[29] The administrator of 1906, M. Spas, also notes that M. Le Conte "does not complain of a lack of labor. In fact, he has an average of 60 to 80 workers who exploit the forest."[30] While in 1901, it seems Le Conte paid 1.20 francs per piece, the 1910 district head, Fillastre, learned that Le Conte paid wages equivalent to 80 centimes per piece plus rice. Thus "labor never failed the *colons*" of Antanambe.[31] Another district administrator of 1911 writes: "Due to his efforts and to the good treatments, due to a scrupulous honesty when it concerns paying natives their salaries, M. Le Conte of Antanambe has managed almost constantly to have the workers he needs."[32]

In addition to his respectful treatment of workers and his "scrupulous honesty" regarding wages (implying the deficient scruples of other concessionaires), workers appreciated Le Conte's piecework system. Presumably Le Conte tried piecework in light of other concessionaires' failure to acquire labor, and in tune with the desires and expectations of Betsimisaraka workers. Contrary to colonial labor ideology of the time, piecework generated a greater valorization for Mananara-Nord than did daily wage work. Piecework enabled peasants to allocate time as they desired to timbering. Le Conte's fair payment appears to have encouraged men to extract rather than incinerate at least a portion of their trees to sell to the concessionaire. Nearly the entire working population of Antanambe village worked for Le Conte, according to one annual report.[33] Piecework enabled Betsimisaraka peasants to earn cash when they desired while maintaining the rituals and tasks of rice cultivation and other agricultural labors, including cash cropping. While the colonial administration pressured Betsimisaraka peasants to enter into wage-labor processes to pay taxes, peasants found it morally and economically imperative to maintain their obligations to the ancestors by continuing to practice *tavy.*

Despite Le Conte's early successes, piecework was not condoned by many of the district administrators. Perhaps envy set in. They complained that Le Conte's wages were too high and that piecework neither encouraged the ethic of regular diligence nor tapped the full market potential of the forest:[34]

> The labor [Le Conte] uses, totally random and on which he can't count for the next day, does not allow him to commit himself too far ahead in time for the delivery of hefty orders of wood, notably that would suit certain houses of England, France, and East Africa. M. Le Conte also had to refuse an important order of rail way ties a few years ago, for he did not have at the moment a sufficient number of native workers wanting to trouble themselves to work to earn up to 5 and 6 francs a day. None earns as much at present.[35]

Betsimisaraka workers' success in obtaining better wages while preserving their autonomy undermined administrative ends. Piecework enabled them to participate in the capitalist market without being "made low" by relinquishing their labor to someone else. By fixing prices to logs or kilos of crops rather than to man-days or -hours, piecework offered peasants a seemingly more egalitarian form of payment than the discrepant wages doled out by foreign concessionaires. It also trained their eyes to the potential exchange value of their ancestral lands, covered with precious and fruit-bearing trees.

CONCLUSION

Land, labor, and unequal relations of power are interlocked categories that have guided how colonial practices have degraded Betsimisaraka people and rain forests. Colonial bosses folded together perceptions of a tropical landscape and its inhabitants as they sought to impose a capitalist-conservationist ethic on Betsimisaraka subjects. Ways of seeing, working, and relating to nature are always evolving. Dominant and subaltern actors represent historical constants; the substance of friction more or less changes. Work, nature, time, and space were not experienced socially as distinct conceptual categories but bled into one another: "killing trees" meant "killing time." Rural Malagasy, for example, might evaluate a worker in terms of hardened feet and walking ability; a French official might reckon labor's value in terms of the number of trees burned or felled and carried away—in this case, it would appear deficient.

As French colonial officials assessed the exploitative value of eastern Madagascar's Grande Forêt, they knew that the valorization of the forest's natural wealth depended on turning Betsimisaraka peasants into workers dependent on wages, a move that necessarily reorganized space and time. The French conception of valorization collided with the Betsimisaraka people's conception of what made work meaningful. Rather than annihilating resistance through a regime of forced labor and taxation, the colonialists contended with Betsimisaraka difference to a certain extent by bending to it. Betsimisaraka peasants have largely spurned subpar wages, unscrupulous bosses, debasement, and the time-discipline of wage work, the latter of which demands a separation from their fields and ancestral obligations. Piecework, in which income derives from fixed prices on natural objects rather than from labor days, has enabled peasants to participate in the cash economy and in their agricultural labors concurrently.

Expatriate conservation and development managers in eastern Madagascar today envision commodity manufacture and ecotourism as means of phasing out "maladaptive" land-use traditions, enriching the countryside, and making peasants more receptive to the conservation's rationale. Ethnographic research suggests, however, that regular wage labor continues to represent for Betsimisaraka workers a structure of constraint and culture of racism comparable to earlier forms of forced labor that obligated workers' time and made them "low."

Figure 1. A logging encampment deep in the Masoala Forest National Park. At each stage in the transport, large encampments exist along the rivers of the national park. At this camp more than one hundred men stage daily expeditions into the forest to search for rosewood or to work the ropes dragging it downstream. The camps are based on crude structures of palm leaves across A-frames and arranged into teams normally associated with a single patron or subcollector. *Photo by Toby Smith/Reportage by Getty Images, August 22, 2009.*

Figure 2. A team of young men drag a rosewood log from the site of felling to the rafting encampments farther downstream. The logs are dragged one meter at a time in teams of four or five for up to six kilometers along the streambeds. The work is extremely physical, and songs are chanted to help with the timing of each movement. Injuries to limbs are extremely common. *Photograph by Toby Smith/ Reportage by Getty Images, August 20, 2009.*

Figure 3. Colonial official transported by *filanjana* in Madagascar, circa 1900. *Courtesy of Foiben-Taosarintanin'i Madagasikara, Antananarivo, Madagascar.*

Figure 4. The crystal ball–making machine at the house in Mananara-Nord. *Photograph by Genese Marie Sodikoff, 2001.*

Figure 5. Betsimisaraka men with pirogue made from a whole tree, circa 1900.

Figure 6. Betsimisaraka woman walking in banana plantation in Mangabe, 1901. *Photograph courtesy of Foiben-Taosarintanin'i Madagasikara, Antananarivo, Madagascar.*

Figure 7. A GCF meeting convened in a village of the Mananara-Nord prefecture. *Photograph by Genese Marie Sodikoff, 2001.*

Figure 8. Raft that ferries vehicles and travelers over a river on Route Nationale 5. *Photograph by Genese Marie Sodikoff, 2000.*

Figure 9. Family posing with newly purchased corrugated iron sheet, foam mattress, and boom box radio in Varary. *Photograph by Genese Marie Sodikoff, 2001.*

Figures 10 and 11. The original Place de l'Indépendence monument, and its transformed appearance in Mananara-Nord. *Photographs by Genese Marie Sodikoff, 2000.*

29
MARS
1947

Figure 12. Man laboring his terraced plot with zebu cattle in Varary. *Photograph by Genese Marie Sodikoff, 2001.*

Figure 13. Woman sowing rice in her *tavy* plot.
Photograph by Genese Marie Sodikoff, 2001.

Figure 14. Offering stone in a *tavy* field in Varary.
Photograph by Genese Marie Sodikoff, 2001.

Figure 15 on right. Checking fishermen's catches at the end of the day in the biosphere marine reserve. *Photograph by Genese Marie Sodikoff, 2001.*

Figures 16. Conservation agents patrolling the national park during the *déguerpissement* of 2001. *Photograph by "Raleva" on behalf of Genese Marie Sodikoff, 2001.*

4 Toward a New Nature

Rank and Value in Conservation Bureaucracy

With the great push toward protecting biodiversity in the late 1980s, conservation representatives coined a neologism, *tontolo iainana* ("the lived-in world"), to translate the concept of "environment," for which the Malagasy language had no good equivalent. The phrase was meant to replace the more familiar term for natural resources, *zavaboary,* which, as the biosphere conservation agent Jafa explained, was usually understood as that which lies "on top of the land, on the surface." *Zavaboary* are gifts bestowed by God (*Zanahary*) to human beings for their consumption. Rather than being an indigenous idea, however, *zavaboary* was most likely also introduced by outsiders, specifically British missionaries, in the early nineteenth century to translate the concept of "Creation." *Tontolo iainana,* in contrast, denotes "everything all around, even in the ground, the land, animals, people, cities, cars," Jafa said. It suggests a more global and secular idea, the idea that humans are elements in an interdependent ecosystem.

In practice, however, I seldom heard Betsimisaraka villagers use either *zavaboary* or *tontolo iainana.* They instead tended to speak of natural things categorically: animals (*biby*), plants (*folera*), trees (*kakazo*), soil/land (*tany*). Betsimisaraka residents of the biosphere reserve commonly heard "tontolo iainana" said in the context of meetings convened in their villages about conservation. *Tontolo iainana* remained a foreign concept, not merely because it synthesized Malagasy words in a new way but because it rang Merina. Since the phrase was in the Merina dialect, it indicated the speaker's ties to outsiders (*vazaha*), as well as an outsider's removed relationship to the land.

Phrases such as *tontolo iainana* were nuggets of abstract knowledge, expressive of a relationship to the natural world that arose from an elevated or distanced perspective, a global vision, that had purchase in the conservation and development sector. In the course of my conversations with conservation agents, they would at times integrate scientific or otherwise foreign-sounding terms into their speech, perhaps trying to impress me with their investment in conservation or commitment to the cause in hopes I would put in a good word for them to the bosses. For example, as I was walking with Sylvestre, the conservation agent, one day in March 2001, he stopped to pluck a leaf from a sapling on the path. He crushed the leaf to extrude the anise-scented sap. "This is *mamitranjetry*, or *Vepris lindriana*," he said, and then crushing a different, lemon-scented leaf, he called it "*tolongoala. Vepris nitida.*"

Since at least the early 1990s, unskilled Malagasy workers in the conservation and development sector have learned the importance of education and educated language for upward mobility. So when I heard conservation agents uttering Latin species names and phrases such as "tontolo iainana" in front of conservation representatives and officials from the Ministry of the Environment, it was clear that these were attempts to get promoted. Conservation agents equivocated between wanting to use the conservation vernacular when they felt their other contributions to the conservation effort were underappreciated and not getting them anywhere with ICDP bosses, and choosing to avoid foreign-sounding words around villagers, since such terms were elitist and tainted by association with *vazaha*, including Merina. The problem of association with outsiders jeopardized conservation agents' standing in their own respective communities, and thus risked eroding the social bonds that ensured their livelihoods when and if they no longer worked for the ICDP.

The conservation bureaucracy and its discourse of science have mediated labor's value in Madagascar and legitimated the logic of rank. In the French colonies, abstract, scientific knowledge about nature, and skills in how to manipulate nature, served as a gauge of a forester's quality and institutional visibility. Historically, certain workers have been positioned in such a way as to make their talents and tasks appear "closer to nature," and therefore less valuable, than others, even though their existence was indispensable to the colonial forest service. Within the neoliberal-era conservation efforts, it was striking how much scientific knowledge and language mattered as a means of moving up the ranks of the ICDP, since what appeared to be the fundamental problem in protecting the national parks remained a shortage of forest guards and extension agents, or unskilled manual labor, relative to the size of the protected area. Scientific expertise, particularly at the higher echelons of the bureaucracy, was abundant, and scientific studies by foreign and national researchers in Madagascar's nature reserves were ongoing and accessible to conservation representatives.

In this chapter, I trace the shifting criteria by which labor within the colonial forest service and its successors has been ranked, drawing out the logical inconsistencies of this value system and the implications of the devaluation of manual labor. I also consider the degree to which subaltern workers have assimilated, mimicked, or rejected the values that promise advancement. Like the criteria of nature's value set against the island's transforming landscapes, the criteria of labor's value within conservation institutions have changed as successive states have articulated their political ideologies in newsprint and on the airwaves. Understanding the viewpoints and practices of conservation agents in the neoliberal era therefore demands a reflection on the lingering impacts of colonial and socialist ideologies on workers' consciousness. Neoliberal conservation and development has negotiated past political rhetorics in search of a new message, in part by resuscitating colonial-era structures, in part by suppressing the categories of "labor" and "peasant" in political discourse, and in part through the decentralization of conservation bureaucracy, which has served to diffuse and obscure who has authority over peasant labor in the rain forest.

CONSERVATION LABOR IN COLONIAL MADAGASCAR

Before France's conquest of Madagascar in 1895, the Merina monarchy claimed ownership of the island's primary forests but allowed localities to exploit natural resources with little intervention. Traditional prohibitions (*fady* or taboos) served to protect certain forest stands from burning or razing (Ramanantsoavina 1966). A year after formal colonization, in 1897, the French state established the first regulations concerning forest conservation for the purpose of capitalist exploitation, including mining, plantation agriculture, and logging by concessionaires who were granted forest parcels from the state (You 1931:406).

Head of the colonial state from 1896 to 1905, Governor-General Gallieni devoted his energies to pacifying insurrections and promoting the *mise en valeur* of the new colony. Ecological damage wrought by the construction of transportation networks alerted Gallieni to the dire economic effects of deforestation. He established a short-lived experimental garden and a forestry school for Malagasy at Nanisana near the capital, where in 1897 he studied which indigenous or exotic tree species could be propagated (Gallieni 1908). He prohibited *tavy* in state-owned forests and decreed in 1900 that concessionaires obtain official permits to exploit state forests.

The first reforestation plantations were created by the colonial military along the border of the eastward road, and reforestation activities continued for the next decade (Louvel 1950:43). The trees were to be used to feed the steam locomotives upon completion of the Tananarive-Côte-Est (TCE) railway, begun in 1901. Imported Indian and Chinese "coolies" began initial construction of the railway until the state made the move to *prestation* labor, a kind of labor tax adopted from the Malagasy labor system of *fanompoana,* in which subjects of the crown received the "honor" of performing unremunerated service to the queen or king (Bouche 1991; Campbell 1988; Feeley-Harnik 1991). In a memoir by Colonel Francis Cornwallis Maude, *Five Years in Madgascar,* the author remarks:

> Naturally, this honor is not always appreciated as perhaps true loyalty would enjoin, although there are a few countries where so much loyalty and even devotion to the Sovereign exist as in Madagascar. But when loyalty takes the shape, as is constantly the case, of carrying vast weights of wood, iron, or stone on raw and bleeding shoulders, along goat tracks (for roads there are none,) through swamps and forests, up and down hills 5,000 feet high, then the additional stimulus of shackles and leg-irons is needed to persuade the poor captured peasant that on the whole he had better accept the "honor," half starved though he must be. If he runs away, he brings punishment on his family and becomes a fugitive and a bushranger; the numerous robber bands are mainly recruited from such runaways. (Maude 2011 cited in the *Spectator* 1895:244)

Reforestation works were left to the state military under Gallieni (Thompson and Adloff 1965). The Departement des Eaux et Forêts (hereafter Eaux et Forêts), established in 1900, had insufficient manpower, staffed as it was by a few Europeans serving nominal advisory roles. The state then abandoned its forest regeneration activities in World War I to concentrate its manpower and resources on the war effort. Meanwhile, private and state loggers, Malagasy farmers, and the railway system consumed vast amounts of timber from the primary forests. At the war's end, Eaux et Forêts focused on recruiting qualified personnel for reforestation and the prevention of brush and *tavy* fires. Former French soldiers sought jobs in the colonies, which offered better pay than in the metropole.

A colonial official, André You, describes the personnel structure of the forest service, a third of which was comprised of former ranked military men. The top tier consisted of officers of the metropolitan cadre of Eaux et Forêts, and subordinate tiers were staffed with men of progressively lesser rank down to the fourth tier, which included French agents of the "fifth class," as well as a special indigenous cadre of agents (You 1931:413–414).

Employees of Eaux et Forêts performed a variety of tasks, including selecting trees to extract, collecting license fees from concessionaires, watching over forest reserves and concessions, enforcing the ban against *tavy*, documenting infractions against the forestry code, arraigning wrongdoers, managing timber extraction for railway sleepers and locomotive fuel, creating tree nurseries, replanting with both fast- and slow-growing species, and monitoring logging activities to determine whether they accorded with the principles of proper forest management (Louvel 1950:43–51; Boussenot 1925:18; Lavauden 1931a).[1] These kinds of activities were considered a form of skilled, scientific labor that could be taught and applied anywhere, such as the Ecole Nationale des Eaux et Forêts in Nancy, France.

In the early 1920s, graduates of the Ecole Nationale des Eaux et Forêts de Nancy arrived in Madagascar to form the superior cadre of Madagascar's Eaux et Forêts and were assisted by a detachment of employees (*préposés* was the term used for forestry employees) from the metropolitan forestry cadre. The inflow after World War I of French and Malagasy foresters trained in France inspired an amount of optimism in Principal

Inspector Griess of the forest service, who found the local European cadre to be "taking on a real worth."[2]

As for the indigenous personnel, Griess felt that the official ranking, which still had not occurred, of former Malagasy soldiers on the forest service staff would stabilize the service's structure and make it easier to implement tasks. Eaux et Forêts sought to devise ways to reforest landscapes with a skeleton crew already overburdened by administrative duties and thus negligent of technical ones.[3] Griess wanted to replace depleted forests with a variety of fast-growing native species. He also thought up a plan, never realized, of creating an obligatory training institution for concessionaires as a way of motivating them to reforest their exhausted woodlots.[4]

When the state in 1927 delimited 353,597 hectares of forest as twelve nature reserves, placed under the control of the Natural History Museum of Paris, it signaled a growing anxiety about the colony's rapidly diminishing rain forest.[5] Henri Humbert, the colonial natural scientist, was instrumental in creating these *réserves naturelles intégrales* (see Dorr 1997).[6] He argued that deforestation left the soil exposed to the "brutal penetration of direct sunlight" and "profoundly disrupted the biological equilibrium" (Humbert 1927:10–11).

Reforestation was a necessary complement to the creation of botanical reserves that since 1927 have legislatively placed greater and greater stores of "autochthonous vegetation" under state control (Humbert 1927: 13).[7] Natural scientists considered the reconstitution of overexploited forests to be less of a priority than the protection of "primitive" forests. Yet abusive woodcutting by concessionaires played just as "nefarious" a role as *tavy* in destroying fragile tropical forests, and reforestation work would provide state and private enterprises the timber reserves they needed, ideally enabling them to avoid further exploitation of primary forests. Colonial foresters established timber plantations of pine, eucalyptus, and native species such as *hintsy*—the "teak of Madagascar"—amidst brushwood and sparse stands of forest along the TCE railway and eastern littoral in 1928 (Louvel 1952:50).

To the consternation of the forest service, central finance administrators continued to funnel a larger proportion of funds into public works projects than into a reforestation program. In doing so, the central government demonstrated an opinion that short-term gains of *mise en valeur* took precedence over the *longue durée* effects of forest conservation. Time and again, the colonial state condoned deforestation for the purpose of bringing resources into capitalist production, while *tavy*, because it incinerated potential exchange value, was vilified. Yet at the same time, the state implicitly encouraged *tavy* production at the sites of public works, where forced laborers had to rely on their kin networks for sustenance at the work camps.

THE PROPER PROFILE OF A FORESTER

The creed of colonial foresters reflected the French empire's overarching labor ideology and policies. Official foresters in Madagascar believed that the colony's ecological stewardship was so important as to warrant the strictest qualifications for the forest service staff. Like other state bureaucracies, Eaux et Forêts provided its staff training in the application of scientific forestry. Graduates of the forestry school in France were recruited by the chiefs of colonial forest services in Africa and Indochina. Colonial officials' assessments of Malagasy nature were guided by a conception of scientific "rationality" that masked the problem of scientific knowledge being overwhelmed by the forces of nature as French foresters died of malaria and dengue fever, or otherwise languished in their rustic outposts while being carried along by Malagasy underlings.

High-ranking forest service officials considered Malagasy intellectually and behaviorally unfit for skilled forestry work but adequate as brute manpower. They believed that specialized forestry tasks could be performed only by agents who had mastered the science of forestry. Meanwhile, commissioned foresters concentrated on measuring tree volume, cataloguing species, marking trees for felling, growing seeds in nurseries, fertilizing leached soils for the plantation of young trees, and the like.

The existence of a low-wage, unskilled workforce in the forest service was scarcely discernable in officials' reports, particularly with regard to Malagasy workers who found employment in the forest service's lower ranks as porters, masons, carpenters, and guardians of supplies. Rural Malagasy had more opportunity for employment in the forest service's railway service obligations—extracting fuel wood from the forests, selectively cutting trees for rail ties, loading and unloading rail cars, and stockpiling and storing reserves of lumber.[8] Eaux et Forêts stationed a large percentage of Malagasy workers at Analamazoatra,[9] felling and supplying timber to fuel locomotives on the TCE railway. These tasks offered Malagasy little opportunity to learn the technical skills of sustainable forest management or to absorb elites' perspectives of peasant practices.

The salaries, backgrounds, and numbers of much of this class of labor were not discussed in forestry reports; their existence is discernable only by inference. Lavauden mentions forest station "ouvriers" only in discussions of their deaths due to plague and of the effort to reconstruct their living quarters in 1929 with durable wood and corrugated sheet metal, rather than soft wood, twine, and thatch; solid materials would presumably prevent pestilent rodents from entering houses.[10] The only clue that the *ouvriers* might be Malagasy lies in the fact that European forest agents were generally called "préposés," and Malagasy agents with official rank were identified as "indigènes" in forest service documents. Like the domestic labor performed by women and girls the world over, the menial labor of *indigènes* was eclipsed by the grand ideas of colonial officials and their strained attempts to achieve them.

By 1924, the number of Malagasy worker days needed by four provincial administrations to manage forests, transport fuel wood from the forest to the rail, and carry out excavation works for the railways totaled 18,873, and in 1925 it totaled 131,422 days.[11] The insufficient manpower of Eaux et Forêts was not due solely to an overstretched budget. After all, Malagasy foresters received a fraction of the salary of Europeans and could have been hired in lieu of Europeans.[12] Rather, the colonial forest service sought employees with a technical knowledge alongside a willingness to perform arduous and exacting work.

It is highly debatable whether forestry was more demanding than the road and rail work relegated to Malagasy forced laborers, or whether rural Malagasy cultivators lacked the floral species knowledge desired by colonial foresters. In the minds of colonial officers, the proper forester was European, ready to carry out the lonely and courageous routine of apprehending *tavy* farmers and unauthorized European and Malagasy lumberjacks and miners exploiting forest land. Reminiscent of the intrepid colonial pioneers of romantic imaginations, the forester would be stationed in a post far from any town and nestled in a shady, wet woods that harbored malaria and the plague. Without a horse or mule, he either had to travel on foot on steep and slippery forest paths or else be carried slowly on a palanquin.[13] The opportunity was less attractive for a commissioned officer than the supervisory roles typically found in *Travaux Publics* positions.

The way in which French officials portrayed the hardships of forestry work from the European's perspective valorized the efforts of Europeans who managed to succeed in the service. The forester's scientific knowledge added cachet. He had to be able to identify species and soil types, to prepare fragile soils for the planting of native and exotic saplings, to know the principles of selective tree felling, and to measure stands of trees. Forestry represented a service to the public good insofar as the afforestation of prairies and plains, the tree plantation in depleted forests, and the preservation of original forests could stem the "disastrous consequences" of "floods, atmospheric disturbances, and . . . all the well known ravages."[14] In addition, certain botanicals could contribute to local and worldwide medicinal needs.[15]

A small contingent of indigenous forest guards who had served in the military did in fact meet the state's criteria for forestry work. As a rule, however, colonial foresters found indigenous forest guards wanting. They did not trust their diligence or loyalty to the service. Plus, there was the problem of possible conflict of interest when a Malagasy guard had to patrol *tavy* plots in the forest. He may have shown leniency, or at least "passivity," with respect to imposing penalties on farmers for their infractions of the forestry code.[16]

Dimpault, an inspector of Eaux et Forêts, voices certainty in his 1928 report that the forest service would have difficulty recruiting Malagasy employees, not only due to the aptitude requirements but also because the work was difficult and unlikely "to tempt young *indigènes* possessing a certain instruction, who will always prefer to take some sedentary job."[17] Lavauden insists in a 1931 report that the forest service "will only recruit young people from the Ecole de Nancy . . . who, despite their general culture and their professional training, can't but very exceptionally offer immediate services. We can only hope to recruit among experienced, ranked officers who are attracted by the lure of the colonial salary. We will not have need for any outside elements."[18]

Although forest officials complained of a sedentary job preference by Malagasy youths over a desire for the rigors of forestry, Eaux et Forêts reported a high attrition rate of European employees in 1930. Lavauden's report explains that in the course of three years, the forest service lost half of its European staff: two were sent back to France for health reasons, two died, one retired, and four had left on vacation or had asked to be reintegrated into the metropolitan forestry service. Among young men arriving from the Institut de Nogent, "one quit the forest service to avoid being fired, another joined the police force, a third gave so many problems that he had to be removed by disciplinary measures, and another, after having submitted, retracted, and resubmitted his resignation, ended up under a serious investigation for dishonesty."[19] Yet Lavauden asserts with pride that no matter the lack of qualified staff, a forestry corps "cannot be improvised." It could not draw indiscriminately on the local inhabitants, be they European or Malagasy, even though local inhabitants appeared far better adapted to life in the forest. In the military culture of the forest service, outward form mattered. Climbing the rungs of the bureaucratic ladder involved a continual assessment of one's performance by superiors. Training for forestry employees followed an apprenticeship model in which young recruits worked alongside seasoned employees.

For Lavauden, the untrained *brigadier* arriving in the colonial forestry corps had to learn a "new nature"—both the nature of disciplined forestry work and the material nature of Malagasy forests. Reiterating the point that "no more than a marine corps can, can a forestry corps be improvised," he argues that the metropolitan forest service was the only acceptable source of labor even though this labor "reservoir" was dwindling in the interwar years. Christian Kull notes that in 1934 Lavauden compares the scant number of foresters, sixty-six, with the national territory of 55 million hectares (Kull 2004:190). For Lavauden, a well-formed personnel was not only a question of material order, but of "moral" order, and his stance on the appropriate qualities of forestry workers suggests a tacit rebuttal to administrative proposals to hire Malagasy men to resolve the labor problem.

Supplying resources for colonial industrial and commercial needs sustained the forest service budget but did not much promulgate the spirit of conservationism among Malagasy laborers. Alongside the problem of forest degradation and insufficient forestry personnel was an increase in international pressure on France to abandon its colonial forced labor policies because they defied the International Labor Organization's 1930 Convention against Forced and Compulsory Labor.

THE "MALGACHIZATION" OF THE FOREST SERVICE

In 1936, France was pressured to retool its public works services to admit the voluntary recruitment of labor with pay. A policy change succeeded the regime of forced labor that had been instituted in Madagascar in the mid-1920s through the Service de la Main-d'Oeuvre des Travaux d'Interêt Général (SMOTIG). SMOTIG had commandeered male agrarian labor into work camps for a period of two years for the construction of railways, roads, and other public spaces (Covell 1987; Sharp 2003; Sodikoff 2005).

During the same year, the state also reorganized the understaffed Eaux et Forêts by adding a greater number of Malagasy employees to the service. The problems of insufficient labor and rising costs within all state departments, alongside a simmering unrest among Malagasy subjects, pushed the state toward "indigenization" of its departments. As Frederick Cooper (1996) argues in his history of African labor in French and British colonies, the recognition that African workers constituted a proletariat deserving of the same rights and treatment as metropolitan workers did not reflect moral compunction turned into policy on the part of French administers. Rather, labor strikes, general protests, and episodes of violence intruded these realities on the consciousness of colonial powers, who were forced to respond.

A set of preliminary report notes by colonial officials in 1936 mentions the reforms: "First, the simplification of the functioning of the service; second, a sensible reduction of staff; third, the replacement of a certain number of European agents by indigenous agents."[20] The incomplete notes mention the possibility of cutting down the staff of European employees to the number present in 1914. These plans showed a pragmatic response to the civil rights protests by Malagasy subjects rather than a sea change in colonial officials' valuation of Malagasy labor. The replacement of European forestry agents with Malagasy suggests that officials had given up on their pursuit of an ideal forest service.

During the 1947 uprising against the French, originating on the east coast, rebels targeted foresters and forest plantations for acts of violence:

> Many forestry posts were destroyed, forest plantations were burned, and a massive increase in fires and *tavy* was noted (Humbert 1949; Deschamps [1960]; Olson 1984; Bertrand and Sourdat 1998). The pot had boiled over, not just in terms of colonial domination, but also specifically in terms of the colonial appropriation of resources (like forests) and repression of natural resource management techniques (like fire)—which for many farmers and herders was their key interaction with the colonial state. (Kull 2004:225)

The aftermath of the Malagasy rebellion in 1947 involved a strict suppression of political activity but a softening of the prohibition against fire for land management (Kull 2004:225). With the *malgachization* (the replacement of Europeans by Malagasy personnel) of the forest service, higher-ranking positions were filled mostly by educated Merina individuals.

In addition to the problem of the Malagasy nationalist movement, and the solution to the labor shortage in the civil service (the hiring of Malagasy), colonial powers in Africa were contending with the specter of soil erosion, of which they had a scary preview during the Depression era when dust bowls bedeviled the United States and devastated agricultural production (Beinart 1984; Anderson 1984). Not until 1943, however, did the French in Madagascar establish a Direction for the Production of Soil to concentrate on increasing agricultural output and stemming soil degradation.[21] In 1948, an Office of Soils Conservation was formed.[22] By the 1950s, soil erosion was the "problem *à la mode*"[23] of the colonial state in Madagascar, as well as in Africa more broadly.

A. Kiener, the principal inspector of Eaux et Forêts, viewed the persistence of deforestation by *tavy*, which induced soil erosion, in the province of Tamatave in 1957 as a "question of authority," a matter of sanctions not being applied. The forest service personnel was still too small; Kiener states: "To date, for a territory equivalent to a sixth of France and which, in addition, carries ⅗ of the island's forests, there is not but three officers, a few engineers of works and a few controllers, around thirty guards for the general service. At present, a guard must surveille on average in his station more that 100,000 hectares of primary forest!"[24]

Ironically, it was not until after independence in 1960, and especially after the May revolution of 1972, that the civil service ballooned to a size that might effectively conserve the eastern forests. But by then ecological conservation was no longer a priority.

Neoliberal conservation unfolded in the context of what was widely held to be a "failed socialist experiment" under Didier Ratsiraka's state. When I was in Madagascar in 1994, a time when most of the island's ICDPs around newly created national parks were just getting off the ground, I gained a sense of the extent to which ICDP workers had embraced certain socialist values articulated by Ratsiraka in political speeches on the radio and television. The revolution had made an impact on Malagasy workers' expectations of the workplace. This was especially notable in light of the neoliberal turn away from revolutionary ideology, such as the vaunting of the proletariat, students, the peasantry, and agricultural self-sufficiency. Replacing these were environmental values, such as biodiversity protection, and global ecological thinking, or the view that environmental problems are interconnected at local, regional, and global scales (*tontolo iainana,* "the lived-in world"). State discourse again fetishized scientific knowledge about the natural world. I got a fleeting insight into the way conservation agents and other manual laborers understood their role during the transitional period of the 1990s, when ICDPs were being introduced into rural areas. When I returned to Madagascar in the summer of 1997, I learned of a work strike that had occurred in December 1996 at the Andasibe-Mantadia Protected Area Complex. In order to grasp the significance of the strike, I will turn to a brief summary of the revolutionary era in Madagascar.

THE SOCIALIST REVOLUTION

In 1972, a movement of students and workers staged a wave of protests against the policies of President Philibert Tsiranana, who finally resigned the presidency in 1972 under increasing pressure to do so. General Gabriel Ramanantsoa assumed power and fulfilled a leftist revolution by liberating political prisoners, lessening French influence in Madagascar, and severing ties with South Africa. By 1975, growing dissent within government regarding the direction of economic policy culminated in the dissolution of government and the rise to power of Colonel Ratsimandrava, who was assassinated shortly thereafter. At this point, Lieutenant Commander Didier Ratsiraka was elected president in a referendum.

Rejecting colonial ties to France and colonial-era policy, Didier Ratsiraka designed his early policies against French models, inducing much foreign capital to pull out of Madagascar (Sharp 1993). Biological anthropologist Patricia Wright, who has done extensive research on lemurs and conservation in Madagascar, writes that the "political climate of Madagascar during the 1970s was not open to western views on science and research, particularly as they concerned species conservation . . . primatologists . . . who had begun their careers on lemur biology had been asked to leave in 1972 and were not welcome to return" (Wright 2008:285).

Ratsiraka sought a radical departure from the colonial past by adopting the socialist rhetoric of contemporary African leaders, advocating isolationism and self-sufficiency, promoting peasant commodity production, and forging ties to the Soviet Bloc countries, China, and North Korea (Covell 1987:29–75). Depending on aid from the Soviet Union for development projects, Ratsiraka's socialism was built upon the promise of a "proletarian revolution" and the devolution of state power to the people. He proposed to organize national development projects through administrative units called *fokontany,* literally meaning "the people." Today, *fokontany* serve to enforce the law and public security at the level of village or hamlet (Laymaire 1975; Gow 1997:413; World Bank 2004:15). His revolution also entailed a surge in civil service employees. By the 1970s and 1980s, the payroll of Eaux et Forêts rose to between seven hundred and nine hundred foresters, a number still insufficient to monitor forest burning (Kull 2004).

By the early 1980s, the state began to "re-open Madagascar to science" (Wright 2008:286). Wright explains that 1981 marks the onset of "a decade of successful negotiations and agreements that have resulted in a much better understanding of the geographic distribution, taxonomy, ecology, anatomy, physiology and behavior of lemurs" (Wright 2008:286). Regarding the conservation of endemic lemurs, in which Dr. Elwyn Simons of the Duke Primate Center in Durham, North Carolina, was fervently interested, captive breeding programs for lemurs outside of Madagascar had been severely jeopardized by scientists' inability to capture and import wild animals from Madagascar since independence, due to the Malagasy state's prohibition. Captive lemur populations in the United States were in dire need of wild genetic stock to survive. By 1982, Ratsiraka had come around, permitting foreign scientists to attend conferences in Madagascar and allowing the capture by North American scientists and removal of wild animals (Wright 2008:288).

While discussions for collaborative efforts for species conservation were under way among scientists in Madagascar and from the United States, Madagascar's national economy was worsening. Real producer prices for agricultural exports had fallen so low that the state could no longer meet its payment obligations to international creditors and donors (Barrett 1994). Economic crisis forced Ratsiraka to turn to support from a variety of foreign sources. Neoliberal reform took the power to determine the rules of the national market away from the state and ceded it to the Bretton Woods Institutions and the World Trade Organization (Edelman 2005). Biodiversity protection became a top policy priority, and donor agencies imposed a different valuation of labor and nature onto entities in charge of forest management and conservation.

THE INTRODUCTION OF INTEGRATED CONSERVATION AND DEVELOPMENT

By and large, neoliberal conservation and development use scientific knowledge (conservation biology, agronomy and agroforestry, and social science) to inform policy. Unlike the case in the past, the public and institutional discourses of conservation and development devoted less breath and ink to the quality of institutional personnel or intransigence of peasants and loggers, for example, than to the interconnected problems of rural poverty and biodiversity loss, to the need to integrate biodiversity protection with revenue generation in rural areas, and to ideas on how to encourage the investment into the conservation effort by rural communities. The new paradigm of "integrated conservation and development" offered "one of our last and most promising hopes for protecting beleaguered natural areas" in spite of the difficulty in reconciling the historically opposed domains of conservation and development (Kremen et al. 1994:395).

As in the past, most high-level positions within the neoliberal conservation bureaucracy, including state and international governmental agencies and domestic and international nongovernmental organizations, have gone to university-educated nationals (the majority of whom are Merina) and expatriates. Lacking postsecondary school education, manual workers have sought to valorize their labor by seeking to accumulate the skills and knowledge that would advance their status in the bureaucracy.

In the early 1990s, forest service employees had to adapt to a new ideological terrain. Now as contractual ICDP employees rather than tenured civil servants, Malagasy conservation workers sought to advance their economic status by tapping into dual rhetorics of value: that which had become fashionable under neoliberalism, and that which was receding from the public discourse as an artifact of socialism. What did this mean in terms of what lower-tier Malagasy workers thought about conservation and development efforts, or their place in the global endeavor?

Lower-tier workers of the Andasibe-Mantadia ICDP offered an oral history of a work strike that had occurred six months prior to my interviews with them in July 1997. I knew these workers from an earlier period of ethnographic fieldwork (1994–1995) in the central-eastern region of Andasibe (Périnet, in French). This was for a masters thesis in international development and social change at Clark University (in Worcester, Massachusetts). My study focused on the tensions between peasants residing near the newly formed Mantadia National Park and ICDP representatives. During long treks between the remote village where I lived with my assistant, Haja, and the town of Andasibe, where we stocked up on provisions, we befriended several of the conservation agents and other manual workers (*ouvriers*) of the ICDP who used to accompany us back to the village to help us carry rice and vegetables. On a brief follow-up visit in 1997, Haja and I learned that many of the conservation agents and *ouvriers* had organized a strike against management, briefly related below.

THE STRIKE AT ANDASIBE

At the time of my initial stay in the Andasibe region (1994–1995), the Special Indri Reserve, the indri being the largest of Madagascar's living primates, and newly created Mantadia National Park—both part of the Andasibe-Mantadia Protected Area Complex—were managed by an ICDP that was run by a North American man, his Haitian-American associate, and a Merina man whose title was national director. The latter was a former forester with Eaux et Forêts. The ICDP's office was an old forest station that stood adjacent to the reserve in the quarter named Analamazoatra ("at the dark forest"), a block of simple houses that had been used for forest station workers and named for its lack of electricity. Analamazoatra lay about a mile's walk from the town of Andasibe, where most of the lower-tier ICDP workers lived. The town of Andasibe had once been a jewel of the colony, possessing a grandiose train station inn with palisander floorboards and bar in the dining room. But when I saw it for the first time in 1994, the station house was run-down and business was slow because the train that had shuttled travelers between the capital and Toamasina was no longer operating.

Most of the Betsimisaraka residents of Andasibe preferred to avoid walking at night along the stretch of road that led from town proper to Analamazoatra. The uninhabited woods along this mile were believed to be haunted by ghosts and malicious spirits (*jiñy*). One night in December 1996, Charles, a guard of an orchid park and employee of an ICDP, was fired for abandoning his post. The orchid park lay in a small corner of the larger Special Indri Reserve. During the previous night, Charles had arrived at his post to find he was alone, his partner having stayed home ill. Engulfed in the pitch-black and the night sounds, the buzz of crickets, rustling, and the plaintive howls of indris from the treetops, Charles could not take it and fled his post.

Charles's former coworkers understood why he had fled that night when his partner did not show up. Yet, at the same time, they faulted him for having left the orchids vulnerable to thieves, who could get a good price for the plants in the black market of endangered species. In the wake of the firing, most of the men were furious with the national director of the ICDP, a Merina man whom the workers called alternately a "shark" (*antsantsa*) or "crocodile" (*voay*). They felt he ought to have issued Charles a warning instead of firing him right off. The right to receive due warning was stipulated in their contract. Charles's dismissal turned out to be the last straw. The thirty-nine manual workers of the ICDP decided to go on strike.

The strikers included unskilled workers (*ouvriers*) and conservation agents, job categories specific to this ICDP (the one in Mananara-Nord did not use the category *ouvrier*). They had already been disheartened in 1993 when they lost tenure with Eaux et Forêts, where most had worked between six and ten years, to instead become contractual workers of an ICDP under the auspices of a North American nonprofit organization that had won the bid to head the project.

The strike of December 1996 had occurred at a point in the life of the ICDP when ANGAP was poised to take over the reins of the project. At that point, the national director would begin to head the project without the Americans. The strikers worried about how ANGAP's takeover would shake up the existing personnel. They felt that their compensation was too low and that the bosses did not respect the terms of the contract with respect to sick leave, medical benefits, training, and the provision of supplies for their jobs, such as rain ponchos and plastic gel sandals for long outings in the rain forest.

In the midst of the work strike, which greatly displeased the management, several of the lower-tier ICDP workers traveled to the capital, Antananarivo, to consult experts at the National Trade Union Syndicate.[25] The *syndicat*, as it was known, was an artifact of the socialist era. It occupied a tiny run-down office in the city. Representatives there had assisted the ICDP workers in formalizing their own union, the first of its kind in Madagascar, comprised of contractual ICDP employees of a foreign-managed project.

At the Andasibe ICDP, and at several others around the island, most of the lower-tier workers were ethnically non-Merina, while the office administrators were Merina. Most of the office workers ("bureaucrats," as the other workers called them disparagingly) were also related to the national director, the so-called crocodile. In interviews with one of the former strikers, he explained his and other employees' many grievances with management, which I chronicle elsewhere (Sodikoff 2007). One man remembered the time when he was forced to guard one of the outhouses used by tourists at the entrance of the Special Indri Reserve because a sink inside the outhouse had been stolen. The conservation agent complained that "he lost dignity" guarding the portable toilet. "It's the dignity of man!!" he exclaimed.

It was not clear to me at the time of the interview whether he had used such revolutionary rhetoric ("the dignity of man"), really deriving from the French revolution, before his experience of the strike and meetings with people at the *syndicat* in Antananarivo. But in using it during our interview in July 1997, he expressed a self-realization of his worth as an ICDP employee.

He and other ICDP workers at Andasibe insisted that their work in the project was worth more than any bureaucrat's since they were "the ones who protected the forest," as well as interacted with villagers of the area. In addition to these attestations of self-worth, the Andasibe workers took pride in their species knowledge of the reserve. One of the men complained to me and Haja about his inability to get promoted and his feeling of being unfairly held back by the national director for reasons that were opaque to him. His job was conservation agent but he was relegated to the role of the "ticket taker" of visitors to the Special Nature Reserve for the Indri. He rued he was no longer to go out into the forest to see and learn about the wildlife: "The guys who work as *agents de conservation* in the forest have a lot of advantages. . . . Me, I'm a conservation agent. Most people who work in the park are. The species *Propithicus* [a lemur] exists here, but I've never seen it! I've never seen it . . ." His disappointment about never having seen one of the myriad lemur species in the reserve reflects a moment in time when the path to workers' self-valorization—particularly in the conservation sector—now depended on the acquisition of ambulatory and optic knowledge about endemic species.

When ANGAP assumed control of the Andasibe ICDP around January 1997, before any of the strikers demands had been addressed, ANGAP authorities laid off the main organizers of the strike, and others involved in the strike received salary cuts. What I found especially interesting in the renditions of the strike as told by the conservation agents and *ouvriers* who stayed on was the way they expressed a conservationist perspective in their accounts, a desire to see the forest protected, to get out in the "field," the deep forest, rather than carry out odd jobs around the office building, and to witness species they had never seen before. These ideas were uttered with apparent unself-conscious enthusiasm, leading me to believe that these few workers appeared to have an authentic interest in the conservation effort. If we look back to the character of the forester that Lavauden had hoped to cultivate in the forest service during the interim between the world wars, we could say that Malagasy ICDP workers appeared to have evolved into the "new nature," one that appreciated Madagascar's "natural heritage" for the nation and the world and sought to protect it from extinction.

After hearing about the strike from ICDP workers in July 1997, I headed to Antananarivo to talk to representatives of the environmental program, including a young North American woman employed by USAID in Madagascar. I mentioned the strike that had occurred in Andasibe seven months earlier. She shrugged nonchalantly and told me that strikes had taken place in all of the island's ICDPs. I later sought to confirm this claim at different ICDP sites, and learned that workers in two other protected areas (Isalo in the south and Fenerive-Est on the east coast) had also struck. Information about the strikes was not readily divulged to me; I had to probe.

The USAID representative expressed surprise at the news that the Andasibe workers had formed a union. She was also interested in the fact that the North American ICDP managers, as well as the national director, had managed to keep the affair hidden from USAID. I supposed that the project bosses at Andasibe did not report the strike to avoid ruining their chances for getting new project contracts in Madagascar or elsewhere, and the national director clearly wanted to keep his post after ANGAP's takeover. The event revealed the extent to which the labor question of conservation was virtually erased from view through institutional practices and individual priorities. Labor-related incidents in ICDPs were omitted from all the reports and program evaluations I could get my hands on at the time. They have also not surfaced in the recently published summary report, *Paradise Lost? Lessons from 25 Years of USAID Environment Programs in Madagascar* (Freudenberger 2010).

Socialist-era rhetoric, exalting the worker as a rights-bearing individual and the peasant as the agent of national self-determination, had by the early 1990s transformed dramatically. The *peasantry* and *proletariat* were no longer the key terms of state discourse. These terms had been suppressed, along with the social aspirations and sense of entitlement they cultivated. In their place effloresced concepts such as "natural heritage," "biodiversity," and "community." And upon retaking the presidency in 1996, after having been unseated for three years by Albert Zafy, Didier Ratsiraka promoted a new national identity, a "Humanist and Ecological Republic," which accorded with the missions of international donors. Labor's value in conservation and development institutions now hinged on a knowledge of species' attributes and locations, as well as on a commitment to conservation and development.

HIERARCHY AND VALUES IN MANANARA-NORD

When I returned to Madagascar in October 2000 for another extended period of fieldwork, this time in Mananara-Nord, I sought to discern whether conservation agents bought into conservation and development. This was always a difficult aim given that who I was (a white North American graduate student) inevitably influenced how people responded to me.

Although Raleva was a "local," what people called a *zanatany* ("child of the land"), he was like Mandresy in that he considered his Christian faith an intrinsic part of his work ethic. More than the other conservation agents, Raleva, I thought, demonstrated an authentic and even fervent commitment to conservation and development. He was the ideal emissary of conservation, save for the fact that his faith lent an air of missionization to his conservation outreach.

Somewhere along the way, Raleva had turned away from the beliefs of his childhood, centered on the relationship to dead ancestors and other spirit entities. His Catholicism and involvement in an association called *Iray Aina* ("One Life") lent him the justification, and the courage, to be *sevère* with peasants. *Iray Aina* was a movement of Christian workers whose life and work was inspired by evangelism. Raleva said its goal "is to evoke a sense of evangelism in our life and work. If you're Christian and sell drugs, for example, is that okay as a Christian?" I shook my head. Raleva nodded in agreement. He listed other examples of how *Iray Aina* sought to make work and the tenets of Christianity coherent: "There are authorities in Madagascar who are Christian but sell our gold, our ore." This was his example of hypocrisy. Raleva found conservation, the protection of nature for the betterment of future generations, to be work that accorded with his Christian beliefs. "In our life today," he continued, "one must follow evangelism and ask oneself 'is this in line with evangelism, these practices?'"

Ironically, Raleva was distrusted to a certain extent by his ICDP bosses, or so he believed, because of his membership in several associations and because of all the volunteer work he did. He thought they thought he had political aspirations. Since the bosses distrusted his motives, they undercompensated him. Raleva had to make arduous outings for the GCF (Gestion Contractualisée des Forêts) initiative, which would allow villagers to sustainably manage forests under the jurisdiction of Eaux et Forêts, so long as they harvested resources according to the guidelines established by the association. If not, the state would take back the forest concession (thereby leaving it vulnerable to illegal extraction, but also making scofflaws subject to penalties). When GCF tasks were added to the list of his routine conservation duties, Raleva had hoped for a raise. He never got one.

Pierrot, a Merina consultant who trained Raleva in the GCF process, eventually intervened on Raleva's behalf, impressed as he was by Raleva's ability and work ethic. He was able to convince the ICDP bosses to increase Raleva's wages. At the time of my arrival, Raleva earned twice the wage of other conservation agents, amounting to 1 million FMG (approximately US$125). But his lack of a baccalaureate prevented him from earning more. The glass ceiling of the conservation and development bureaucracy again recalled the situation of indigenous employees in the colonial forest service, whose ambitions were stymied by institutionalized definitions of expert knowledge.

When I accompanied Raleva on a long trek into villages bordering the core of the biosphere reserve in February 2001, my presence on the mountain trails always attracted a lot of interest, as I mentioned earlier. People were mostly curious and surprised to see a foreigner so far from the main road or Mananara-ville, but on a rare occasion I would get a more aggressive reaction. As I walked behind Raleva in one village, a drunken man delighted himself by laughing at my lameness. He started to perform a clownish pantomime of me, giving me the French *bises* (kisses) on my cheeks. Raleva smiled sternly, irritated, but I shrugged it off. We continued. In another village, another man, also amused by my limping, raced into his house to grab a camera (a rare possession) and pretended to take photos of me. Raleva was offended and at this point turned to an elderly man, in earshot of the man with the camera, and lectured that the other was acting disrespectfully and lessening the chance that tourists would want to come to these parts. No inflow of tourist money here, he implied. The old man listened and nodded soberly while the younger man grew serious, castigated. Raleva disapproved of vulgar behavior, like making fun of an ailing *vazaha*. Respecting custom (*fômba*) by speaking to the elder rather than scolding the younger man directly, he was an effective communicator, an insider, a Betsimisaraka workingman, subtly introducing "modern" ways to the village. His righteousness and patience were inspired by his religion and his approval of the subjective transformation that "development" entailed.

We hiked for hours, long after my knees had given out. Darkness fell. Out of pity, Raleva had taken my heavy backpack on top of his own. The trail had become a vertical rock stairway. The temperature had dropped steeply, and the air shrilled with cicadas. My flashlight spotted a sign on a tree: "Warning Forest Territory of Savarandrano." Impossibly, it seemed, we were nearly at our destination, and I was interested to see that the villagers had already staked their claim to the surrounding forest before officializing any GCF contract. When we arrived, finally, I collapsed on the cot in the wood-plank house reserved for visitors. Raleva secretly arranged for an elderly woman to come to the house to massage my knees with a healing balm.

At dawn, we stepped out of the house into a blanket of low-lying mist. The village was perched above a remnant forest pulsating with the croons of indri lemurs. Raleva's meeting was off to a late start, which annoyed him. Although he had alerted the villagers to his arrival days earlier, the means of communication here were unreliable. Messages were conveyed on footpaths by "word of mouth" (*ambava*). If a messenger's destination fell short of the mark, he or she relayed it to the next passerby. If a message reached the correct village between December and May, chances were that most villagers would be scattered among their rice fields in the hills, their *tavy,* working and sleeping there. Someone would have to blow the *bakôra,* a trumpet-shell horn, signaling a return to the village for an important meeting.

The meeting took place in the schoolhouse, attended by about twenty men and eleven women, many clutching infants. The attendees sat on dilapidated school benches, and children crowded the doorway to peer in. Finally, at 10 AM, Raleva began. He introduced me at length, then himself. Then he proceeded to explain the purpose of the GCF association. "Using up the forest will leave you with nothing," he said. He pointed at one man: "If he goes into the forest," then pointing at another, "and he goes, and Madame over here leaves her cow in the forest, everyone doing what they want, the forest is ruined."

After his lesson in the "tragedy of the commons," Raleva explained that this forest was not part of the national park. If the community concession was not managed sustainably, there would be no contract renewal, and Eaux et Forêts would reclaim jurisdiction over the forest. The meeting droned on, eyes half-mast and faces resting on desktops, as Raleva read the entire community forest management statute, written in the Merina dialect. He translated the words into the Betsimisaraka vernacular since most villagers cannot understand Merina. The villagers named their association SAMIA ("all of us"), from *Savarandrano Mitsinjo Ho Avy* ("Savarandrano Foresees the Future"), a name that self-consciously adopted the conservation ethos, perhaps pandering to the authorities. The GCF initiative was popular, and peasants were eager to gain use rights to local forest concessions. One of the rules stipulated that SAMIA was forbidden from affiliating with a political party; it must remain neutral.

Raleva was there to secure a conservation contract with the villagers. The legalization of the association required paperwork to have it formally registered, but it was difficult to get villagers to sign their names to documents. They were more comfortable with verbal contracts, he explained. Most were illiterate so were wary of signing paper and did so reluctantly, but the association officers finally did. Then came the barrage of distrustful questions at the end of the meeting, the likes of which Raleva could never seem to put to rest: Were the association members signing a contract with the biosphere? Was this forest owned by the biosphere? If a member broke the rules of the contract, would the biosphere take away their forest? Because decentralized conservation involves an array of foreign and state interventions, it made the conservation emissaries seem deceitful, as if glossing over outsiders' real motive of expropriating peasants' land. Raleva's explanations failed to fully convince his audience that he was in fact not an Eaux et Forêts functionary, that he was not a state agent, that the biosphere was a separate entity all together.

Raleva's umbrage at being misidentified as a state agent reminded me of my frustration as a Peace Corps volunteer in the Comoros (Anjouan island, 1989–1991), where the volunteers were often asked by Comorians if we were CIA operatives. We all found it offensive yet understood that the association people made was valid. The conservation agents took offense at being associated with Eaux et Forêts, and did not relish their association with *vazaha,* but all this was complicated further by their involvement in the forest sweeps, called *déguerpissements,* during which conservation agents paired up with gendarmes to rout scofflaws out of the national park, as I recount later. During these sweeps, any strengthening bonds of trust between villagers and biosphere workers through the popular GCF initiative, as well as dam building for irrigation, school repairs, veterinary services, and so on, were weakened.

By 2000, had Malagasy conservation agents, at least in Mananara-Nord, adopted the conservationist sensibility? In the schema of labor value in which species knowledge and the commitment to conservation ranked highly, one might have selected Raleva as the most valuable conservation agent of the biosphere reserve. He appeared to have assimilated conservation values and the ideology of progress. Yet he felt obstructed in his ascent by the distrust of the national director and the Dutch chief consultant. As I got to know Raleva, I could not understand why he was held back, why he was not appropriately rewarded for his commitment to the cause.

Raleva felt that conservation was a means to improving the material conditions and "moral character" of Betsimisaraka peasants in Mananara-Nord. "Peasants know a lot about the landscape," he said to me in February 2001. "Much of what people know, they can't explain. They just know it inside, from a lifetime." Although he understood and respected the emplaced knowledge of peasants, possessing it himself, he saw a need for social transformation. He had ideas about civil behavior, about the ethics of social intercourse, as was apparent in his interactions with subsistence farmers. Although he was not the only self-identified Christian in the conservation crew (there were three), Raleva's ease with expatriate and Merina consultants and administrators, and his fluency in the languages of planners and peasants, made him in many ways the ideal conservation agent. He was able to grasp and empathize with the perspectives of both the peasant and the metropolitan.

Conservation agents like Raleva were the "watching eyes" (*mpitara maso*) of the biosphere reserve. They had authority to enforce conservation rules and occasionally did so with gendarmes at their sides, but they were not civil servants. They saw themselves for the most part as *zanatany* and possessed the ambulatory knowledge of long intimacy with the rain forest and village settlements but were estranged from the standpoint of peasants by virtue of their salary, tasks, and mission.

CONCLUSION

Bureaucratic hierarchy is a semiotic representation of value having spatio-temporal existence; its verticality corresponds to social assessments of what is "backwards" and "modern" (see Salmond 1982; Nederveen Pieterse 2002). Ranks within colonial bureaucracies were meant to reflect stages of social development, the higher levels reserved for ostensibly more civilized persons. In Madagascar, the highest-ranking positions of the colonial forest service were reserved for Europeans educated in scientific forestry, and the lowly ranks were filled by Malagasy *indigènes* who knew the landscape intimately but lacked a certain lexicon and knowledge.

In the neoliberal era of conservation and development, European value systems were resuscitated as conservation and development planners validated a dominant historical narrative. In this narrative, conservation successes have been attributed to scientific and technocratic knowledge (intellectual, expatriate labor), while conservation failures have been attributed to the lack of peasants' schooling in matters related to environmental degradation. For Malagasy workers in the conservation and development sector, moving upward has demanded the acquisition of a certain set of social values about, among many things, the proper human-environment relationship and the proper way to make a living. Historical and ethnographic data make visible the blind spots of conservation and forestry institutions, whose hierarchies reward workers in proportion to their degree of alienation from the wilderness and society they are meant to police and transform. What I suggest is that the contemporary division of labor in Madagascar mirrors an evolutionary schema with deep roots in European thought, and it has played an important role in creating the reality of species endangerment.

Le temps n'est que l'activité de l'espace.

—ELSA TRIOLET, *Le Grand Jamais*

5 Contracting Space

Making Deals in a Global Hot Spot

The first time I saw Mananara-Nord was in July of 1999, when I made an exploratory visit there with Haja. We went overland from Toamasina, where about fifteen passengers piled into the back of a *camion-brousse,* a vehicle that the Lonely Planet guidebook describes as an "army style truck fitted with a bench or seats down each side . . . used for particularly long or rough journeys, which you may well wish you had never begun" (Andrew et al. 2008:284). We sat atop bags of used clothes and rice, and on other people's legs and elbows, as a torrential rain pummeled the canvas cover, making the interior stifling, and at nightfall pitch-black. Once we passed the river town of Soanierana-Ivongo, Route Nationale 5 became a narrow dirt road. Thirty-six miles separated that town from Mananara-Nord, but the rocks that bulged from the road's surface, the deep mud slicks, ramshackle bridges, and portions of soft, sandy beach extended those miles into twenty-five hours of driving time. Route Nationale 5 had been neglected for years—at least, people say, since the early years of President Albert Zafy's term (1993–1996), a brief and hopeful interlude to Didier Ratsiraka's hold on the state since 1975 that ended with Zafy's impeachment.

There are alternative theories as to why such a major route had been left to deteriorate. Some say Ratsiraka ignored road repairs in retaliation for tepid support among citizens of the northeast coast during the 1996 elections. Others say that the Chinese merchant families who dominate the maritime shipment of cloves, vanilla, and fruit from Mananara-Nord to the port of Toamasina want to stave off competition by people with trucks. So, people say, the Chinese made deals with Ratsiraka's ministers to keep the road nearly impassable and their boats in business.

The road skirted the coast—to the east, the Antongil Bay; to the west, a vista of ravinala palms and green rolling mountains. It cut through villages of ravinala-leaf and plank houses, as well as occasional herds of zebu cattle being corralled to new pastures. Northward, the mouths of seven rivers, too wide for bridges, required wooden rafts to ferry vehicles across. Moving northward, the vegetation began to thicken, and the road hugged the shoreline. Just after we crossed the seventh river, in the village of Anove, a sign staked on the side of the road announced our entry into the UNESCO's Mananara-Nord Biosphere Reserve. This was the first clue we were entering a new space being designed, in Arturo Escobar's (1997:210) words, to "deliver nature from the grip of destructive practices and establish in its stead a conservationist culture."

Planners of neoliberal conservation pinned hopes on future revenue from the capitalization of intact nature, a form of "reprimarization," or "a return to a reliance on primary export products," which, according to Fernando Coronil, has reauthorized metropolitan control over postcolonial societies (Coronil 2000:363). In Mananara-Nord, the production of cloves and vanilla, conventional objects of reprimarization, competed with a reliance on the emergent form of reprimarization, or the commodification of whole ecosystems through conservation and ecotourism, as well as through the conversion of rain forests into rent-earning research sites for scientists. Breaching the stage of conventional manufacture, the nonextractive production of rain forest value enacts a "compression" of time and space.

David Harvey's (1990) concept of "time-space compression"—an outcome of the evolution of capitalism which today includes the simultaneity of the new information technologies, high speed transportation, transnational production processes, and the accelerating turnover time of production and the circulation of exchange—describes a transformation of the subjective experience of the world. As Noel Castree (2009:35) explains, "people register cognitively and emotionally the increased pace of change and the enhanced capacity of far-distant places to impact on their own lives instantaneously." Conservation proponents express a low-grade anxiety at the acceleration of habitat loss and increasing pace of species extinctions due largely to the activities of populations at the global economic peripheries. (The problem of overconsumption in the global North is acknowledged but deemed too resistant to change.) But how is time-space compression experienced phenomenologically and mentally in biodiversity hot spots like eastern Madagascar? In Madagascar, did Western declensionist narratives about biodiversity loss and worsening poverty transform *tavy* farmers' perception of time and space? Were residents of the biosphere reserve assimilating the sense of anxiety over species and habitat loss? Were they acquiring a globalist sensibility, a desire to protect rain forest as part of "our common heritage" (Breidenbach and Nyíri 2007)?

The answer, as far as I could see approximately one decade after the creation of the biosphere reserve and ICDP, was that yes, residents sensed an alteration of time-space due to the influx of outsiders into Mananara-Nord, and that no, they did not concede that *tavy* was the cause of this compression or that the rain forest was part of the world's heritage. They were more likely to blame the national park for their feeling pinched or hemmed in, unable to clear new land for crops and therefore unable to extend their line of descent by bequeathing new land to their offspring. Added to this in 2000 was another dimension of globalization that appeared to be rapidly accelerating the pace of social life. The main driver of this energy was cash cropping, particularly the production and exchange of cloves and vanilla.

Cash cropping was one enduring instance of "colonial globalization," a nodal point at which Madagascar had been integrated into the global economy and remained so forty years after independence (Coronil 2000: 363). In eastern Madagascar, peasants had begun voluntarily cultivating cloves under the colonial state as a means of securing land tenure. Mananara-Nord was an important region for cash crops in Madagascar, and through the production of cloves, vanilla, and coffee—colonial crops—it had by the late 1990s emerged as one of the national epicenters of crop wealth, and a space of rapid development in Mananara-ville. A bumper clove crop in 2000 was due to the efflorescence that year of the region's cloves trees, which do so only every three to four years. This event coincided with the increased demand from Indonesia after the collapse of its currency, the devastating forest fires of the late 1990s, and Indonesian growers' neglect of their clove crops (Landau 1999). Madagascar had quickly become a major supplier to Indonesia, where clove cigarettes are widely consumed, and to the rest of the globe.

The majority of villagers with land participated, including the conservation and development agents based in villages of the biosphere reserve. I had heard, but none of them dared to confirm, that the conservation agents had formed an exclusive cash-crop cooperative, not allowing other ICDP employees, such as the administrative personnel, office guardians, or members of the development component, to join. Since no one admitted the existence of the "club," I could not plumb the details. But I had seen the clove trees that lined the buffer zone of the national park near Varary on my tour through the forest edge with Sylvestre. At the time, he explained that conservation agents had planted the cloves years ago, and agents shared the revenue from their harvest.

Expert consultants' reports about the effects of clove production on the biosphere reserve suggested that it complemented, rather than undermined, the protection of biodiversity. If the island's clove corridor remained so profitable, cash cropping could encourage peasants to lessen their reliance on *tavy,* a process that ecotourism development had not yet been able to accomplish. Contrary to implicit assumptions that biodiversity conservation necessitates a certain amount of repression to eliminate erosive forms of production from protected spaces, in Madagascar, the cultivation of economic trees and vines had come to be seen in the early 2000s as environmentally advantageous. However, with the international increase in crop prices, there were indications that the international price boom for cloves and vanilla carried the risk of eventually impinging on the primary rain forest due to the influx of *mpiavy* (migrants, newcomers) who also sought scarce land to plant their own cash crops.

Cash cropping was such an important activity to the region that it was impossible to separate out the effects of conservation interventions on rural people's thinking about the profitable exploitation of land and what species were of utmost value to people's lives. This chapter examines the social relations of cash-crop production and exchange in Mananara-Nord, which were key elements of Mananara-Nord's geographical identity. The forms of value produced through verbal contracts of exchange between producers and collectors, the political exploitation of the value of crops by officials and campaigners for the presidential election, and the revisions of colonialism that emerged during the crop boom highlight the obstacles faced by conservation representatives trying to introduce a novel philosophy of nature's value into rural Mananara-Nord society. I explore different forms of relationality inherent in cash cropping: the relationship between outward activity and inner feeling; the differential power of crop collectors, peasant producers, and state officials; the unstable and opaque collaboration between state representatives and donor representatives; the markers and meanings of insiderness and outsiderness; and the relationship between the colonial past and the globalized present.

Regarding the first point, the relationship between outward activity and inner feeling, David Harvey's "time-space compression," as well as Raymond Williams's "structure of feeling," referring to historically and place-specific habits of thought and practice, offer concepts with which to make sense of how globalization and globalism—specifically, the principle that tropical nature is humanity's common heritage—affect rural populations on the brinks of the world economy. Ethnographic accounts offer insight into the geographical particulars of these trends.

The spatiotemporality and mood of cash cropping in Mananara-Nord were imbued with fledgling attempts by external entities to diffuse conservationism and protect primary habitats there. The latter entailed arresting the degradation of primary rain forest and the coral reef system by educating rural Betsimisaraka people, as well as distant metropolitans, to appreciate nature's intrinsic value and to support the expansion of wilderness. This was a more plodding path toward wealth creation than cash cropping, so the sustainability of the biosphere reserve depended on market-based activities that went on outside the boundaries of the rain forest. Cash cropping put no direct pressure on rain forest and marine resources in the early 2000s, but like conservation, ensured that colonial structures remained palpable realities. The ICDP fit within a layered colonial history, where the governance of land, bodies, and mentalities established enduring "racialized relations of allocations and appropriations" (Stoler 2008:193).

COLONIAL GLOBALIZATION IN MANANARA-NORD

With neoliberal conservation and development in the mid-1980s, decentralization transferred much of state authority over protected areas to Western metropoles, such as Washington, D.C., and Paris. This was the case with the Mananara-Nord Biosphere Reserve during its management by UNESCO. Although I had made contact with biosphere staff the previous year, I had failed to get permission for my ethnographic study from UNESCO in Paris. The United Nations office in Paris claimed jurisdiction over all research conducted in the biosphere reserve. Since villages were within its boundaries, UNESCO's permission was also required for at-home interviews with biosphere employees. The rationale might have made more sense if my research would interfere with employees' work hours, or if I was entering the national park without authorization, but I had no intention of intruding, interrupting, or following conservation agents into restricted territory without permission. It was a strange kind of authority, UNESCO Paris, dictating social science research protocol. Yet it was also a means for the head of the ICDP, the Dutch technical consultant, to assess whether or not I was trustworthy, a matter put into doubt when I had mentioned an affiliation with a former minister of education who was a Betsimisaraka anthropologist who had done his research in his natal territory of Mananara-Nord.

This former minister and the ICDP's chief technical advisor, a Dutch ecologist who represented UNESCO, had a tense and unfriendly acquaintance, the former calling for more transparency in the biosphere budget and resenting the inflated salaries of expatriates who, in the view of the ex-minister and other Betsimisaraka residents of Mananara-Nord, lacked knowledge of the region and possessed no more administrative expertise than many Malagasy candidates. The chief technical advisor, in turn, distrusted the ex-minister's designs on the biosphere. I was told that after having conducted the early studies in the newly designated biosphere reserve for the World Wide Fund for Nature (WWF), the ex-minister had hoped to obtain directorship of the biosphere project under UNESCO. The chief technical advisor believed that the ex-minister used the biosphere project as a political tool to enhance his status among Mananara-Nord constituents. He learned that the ex-minister, after not becoming director of the biosphere project, spread negative propaganda among Mananara-Nord residents to stir up protest against the biosphere. To avoid getting involved in these politics, I decided to hold off on my main study until I received UNESCO's permission, and instead visited the villages of the biosphere reserve to learn about clove production and processing, a topic that apparently needed no approval from UNESCO.

Cash cropping was big business in Mananara-Nord in the early 2000s, and its cyclical culture, being less regular and bringing in large amounts of scarce cash, overwhelmed that of the *tavy* economy. As a form of primitive accumulation in which peasant producers sold their crops to crop collectors, who in turn sold them at a profit to export companies, cash cropping in the early 2000s had evolved from its early-twentieth-century form, when colonial settlers established plantations, vanilleries, timber concessions, and mines in the rain forest, with Malagasy labor earning low wages or doing piecework. Not only did colonial settlers clear rain forest to establish plantations, but so did Tsimihety and Betsimisaraka peasants, who learned that the state would recognize their property claims when land contained cash crops (Fanony 1975). If contemporary exchange relations gave any indication of what they were like a century earlier, then peasant producers then, as today, were keenly aware of the unfair terms of wage-labor exchange (as archival documents suggest), yet were somewhat more tolerant of the terms of piecework, such as in the exchange of cash and bags of crops. In the early 2000s, peasant producers openly offered their assessments of the trustworthiness of certain collectors, and if they trusted a collector, they tended to contract with him every season. And collectors approached exchange with peasant producers opportunistically and, in my view, with disdain for, or at least condescension toward, peasants' gullibility.

The combination of conservation and ecotourism intervention, long maligned by peasants, and cash cropping, which promised accumulation but also drew in outsiders, produced the spatiotemporality of globalization in this particular hot spot, so important to the international effort to protect biodiversity. Mananara-Nord had become a salvage frontier for global capitalists, and a brink of extinction for global conservationists. Many Betsimisaraka peasants recognized the harmful effects of deforestation, caused by *tavy*, colonial-era cash-crop production, and past and present timbering. It caused flooding during the rainy season, a depletion of birds, lemurs, and wild boars, a shrinkage of panicles on the rice stalks, and rising soil acidity indicated by the abundant growth of fern (*ohotra*, Betsim., *Sitcheris flagellaris*, Lat.) and cogongrass (*tenina*, Betsimisaraka). One elderly man, a resident of the village of Anteviabe ("at the place of abundant forest clearance"), remembered a time when "one *kapoaka* of grain used to yield a ton of rice at harvest." A *kapoaka* is a standard measure used in Madagascar, the size of an empty five-ounce can of condensed milk. Planting a can of seed brought ample returns. If "development" had appeared to correlate with deforestation in the past because the more developed and urban areas were cleared spaces, Mananara-Nord residents, according to Raleva, the conservation agent, were beginning to revise their interpretation. Raleva said to me one day in February 2001,

> People here say that the disappearance of forest signifies development. People that live outside of the forest are *vazaha* who have money and nice things. But people are now noticing newcomers from Vavatenina, Fenerive, and Mandritsara, coming here to find work in cloves or on the land. And they are from places where there's no more forest. So people are realizing that the disappearance of forest in the development equation doesn't add up. Without forest, the water is out of control, there's no wood for construction, no natural pharmacy, and so on. (Field notes 2/28/2001)

Raleva's analysis seemed to be tinged with optimism that Mananara-Nord inhabitants would be more inclined to embrace conservation with the inflow of *mpiavy* to the area. In my view, residents' concerns about the future were instead temporarily allayed or mitigated by the potential for rapid accumulation through cash cropping. Cash was in turn traded for status. An important medium of value in rural societies of Madagascar was "familial relations" (*fihavanaña*), or a relationship of mutual loyalty and care, emulating kinship, established between oneself and unrelated others. The wealth earned from the sale of crops was quickly converted into social bonds, as men frequently indulged in ostentatious and reckless spending with crowds of people.

When I returned, again overland, to Mananara-Nord in October 2000 to begin my research, the region had become a jubilant and raucous cash-crop frontier, a magnet for streams of migrants, capitalists, and thieves. The scent of cloves saturated the air, and Route Nationale 5 was now swarming with flatbed trucks moving bags of product as best they could over the mud pits and rickety bridges.

Clove trees (*jirofo*, from the French *girofle*) effloresce every three to five years and tend to "rest" (*malaka*) in between time. In 2000, people poured into the region to make their fortunes in "black gold" (*vôla-maintiñy*), which referred specifically to the high-quality vanilla grown there, although the term might well have also applied to the blackened dried cloves and roasted coffee beans of Mananara-Nord. Throughout the season, the wholesale price for cloves rose—from 15,000 FMG (about US$2.50), to 20,000 FMG, to 25,000 FMG—up to 40,000 FMG (about US$7) per kilo by the time most of the crop was bought up around

March, a huge price hike compared to 750–1,500 FMG per kilo in 1995 (Brown 1999:204).

Villagers sat in their courtyards in front of reed mats piled with mounds of red and yellow flower buds still attached to stems. Repetitively rolling clusters of buds against the palm of one hand, people snapped the buds off their cymes, an activity called "doing the clawing" (*mañano lagriffe,* from the French *la griffe,* or "claw") but more accurately described as "doing the snapping off." "Hello, what's new?" went the greetings then. "Doing the snapping off, ééé!"

The cash-crop season was a time when peasants' acquisition of quick wealth exposed the enduring stark disparities of wealth in the region overall—that is, the novelty of earning a lot of cash from cloves and vanilla emphasized the depth of rural poverty present most of the time. The surplus advantage of collectors and exporters over peasant producers exposed the vulnerabilities of producers in the global cash-crop market. Malagasy politicians exploited the colonial past by casting blame for the nation's economic woes on ethnic outsiders, including Europeans. The manipulation of state symbols and the revision of history around the subject of cash cropping was especially current during the presidential election campaign in December 2001.

Cloves and vanilla were sold by the kilo, which let peasants participate in the market while ostensibly protecting the autonomy of their labor power. The harvest and processing of cloves overlapped with the period when vanilla flowers had to be hand-pollinated on the vine. By June, the scent of vanilla reigned as the oily, cured pods dried under the sun and the protective gaze of their owners. Cured, or "cooked" (*masaka*) vanilla (as opposed to less expensive raw, green vanilla) could capture as high as 600,000 FMG (approximately US$95) per kilo in 2001. Vanilla's rising value began to make the labor intensiveness of its cultivation worth the while. Cultivating vanilla, solitary work, forced one to keep close to the village when the flowers bloomed on the vines, because they needed to be hand-pollinated. Also, vanilla owners needed to keep an eye out for strangers on isolated footpaths who might steal whole vines already heavy with beans. Many people denied owning vanilla plants for fear of thieves. In contrast, clove trees stood out in the open, their yellowing leaves evident in panoramic views of the river valley and mountains. People knew how many clove trees their neighbors owned. A fussy plant, vanilla required a perfect mix of soil and light, as well as patience and a delicate touch. Curing vanilla beans also required expertise. It involved weeks of sweating the beans and wrapping them, careful to avoid rot.

Processing clove flowers occasioned a lot of socializing and visiting among villages in part because it required extra hands. The sociality of clove work enlivened villages in the evening and brought people together to visit, drink, dance, and romance. Clove work was all-consuming. Absorbed by it, growers would forgo their paddy rice and need to buy rice in March and April before the *tavy* fields ripened. A portion of clove wealth was almost immediately consumed in rice, manufactured goods, and drinking and dancing in the restaurant-bars of Mananara-ville.

Villagers of Mananara-Nord had assimilated the colonial crops into their marriage customs and into the subsistence economy. Clove trees were markers of status in marriage negotiations. In the early 2000s, they were an important part of the *orimbato* ("laying down a rock"), or bride price. The *orimbato* consisted of a series of negotiations between the prospective groom and his parents, and the prospective bride's parents, as well as presentations of decorative gifts (*saroño*, including clothes or jewelry) by the young man to the young woman. The groom would visit the bride's family to ask for "a human nursery—not a banana nursery—but a human nursery." The practice of *orimbato* ensures the woman's family that she will be provided for and well treated by the groom's kin. Like the clove trees presented to the woman's family, the marriage will endure and be fruitful.

While villagers processed their cloves, collectors, who were either residents of Mananara-ville or itinerant collectors from more distant towns, visited the rural villages in hopes of securing contracts with individual growers to buy their crops. The exchange relationship between peasant producer and crop collector explicitly recalled the colonial era, as collectors were called "vazaha" for the period during which they played the collector role. The term did not necessarily impute Europeanness to the collector; rather, it signified peasants' acknowledgment of and respect for the greater economic power of the collector.

THE VERBAL CONTRACT

Many peasant growers stored their cloves in their village houses waiting for the per kilo price to rise, and others waited for collectors to come to the village with better prices than offered by shopkeepers in Sandrakatsy. Individual collectors took out loans from the big export companies in town, ventured into the distant countryside (distant enough to disincline peasants to lug their produce into town), and secured contracts with growers before their crops were harvested and processed. Many frequented the same villages every year, where they had established trusting relationships with growers. Even if the relationship between producer and collector was apparently trustworthy, peasants were frequently duped by weighted scales that were brought into the villages or used in the warehouses of the big export companies.

Selling to itinerant collectors saved growers the arduous trip into town, hauling their heavy bags of dried cloves or vanilla to the road, and then finding a bush taxi to take them to the merchants or export firms in town. Intermediary collectors also had the trouble of transporting the crops from village into town, but typically they hired porters to help them, and often owned four-wheel-drive vehicles for the road trip. Not all did, however. Collectors then took the cloves in baskets or plastic sacks to the *Sinoa* (Chinese) stores of Sandrakatsy or Mananara-ville. Merchants in turn sold the produce at a lower price to the big export companies, or, if they were owners of export companies (as were some of the *Sinoa* families in Mananara-Nord), they shipped the crops via Chinese shipping companies to Toamasina. The large export companies with representation in Mananara-ville included Ramanandraibe, a Merina-owned company with headquarters in Antananarivo; Madagascar Cloves; Nim Tack, a Chinese merchant residing in Mananara-ville; CEVOI, a foreign-Malagasy vanilla export company based in Toamasina; Fakra, an Indian-owned company; French-owned Des Landre; and the Betsimisaraka-owned Sieb, based in Mananara-ville.

In October 2000, I accompanied a collector, Jino, a *métis-Chinois,* the term used to describe an assimilated Chinese who spoke Betsimisaraka like a mother tongue and who may or may not (usually not) have Betsimisaraka kin. Jino, who was in his mid-thirties, spent his time between the city of Toamasina and Mananara-Nord trading in cloves, vanilla, or crystal. He was circulating in his pickup truck around roadside villages, unfolding huge wads of cash in front of villagers as he struck deals with growers to sell their crops to him. Villagers greeted him, as they did other collectors, with “mbôla tsara, vazaha!” (“greetings, stranger/outsider!”). Calling Chinese and other Malagasy collectors *vazaha* temporarily conveyed camaraderie and respect, an ironic performance of colonial relations. As far back as residents of the biosphere reserve could recall, since around the 1940s, Chinese, and not French, had been the collectors of Mananara-Nord.

Not only did the memories of colonialism cling tenaciously to the Malagasy conception of *vazaha,* but colonial hierarchies were also re-instantiated in contemporary interactions between peasants and capital. Jino projected grandiose self-confidence and pretended an intimacy with the villagers that bordered on condescension. He relished the role of the *vazaha,* commanding villagers to serve us fruit and cups of coffee sweetened with cane juice. Nevertheless, villagers claimed that the collectors "inspired trust" (*mahafatoky*) without specifying whether they trusted Chinese, Betsimisaraka, and Merina collectors equally (see Brown 1999).

During the clove harvest many men enjoyed "showing off their money" (*mitera manambola*) with excessive and ostentatious spending. In late 2001, growers whose trees refused to "rest" after last year's bounty continued to enjoy rising per-kilo prices. Although the clove crop was leaner during this season than in the year previous, it could still make fortunes. Farmers with spare sacks of dried cloves from the previous year, or with trees that bloomed off-cycle, earned at least 40,000 FMG (about US$7) per kilo. Vanilla prices had continued to rise even higher than 600,000 FMG (approximately US$95) per kilo of dried, high-quality beans.

The marketplace in Mananara-Nord, ever expanding with the influx of migrants eager to sell things to peasant producers, was stocked with a growing variety of temptations for the absolutely poor, such as cheaply-made-in-China clothes and pinup calendars, as well as more expensive items such as corrugated sheet metal for rooftops (*taoliñy,* Fr. *tôle*), battery-powered Sony boom boxes (*sôny*), mountain bikes (*bisikilety,* Fr. *bicyclette*), foam mattresses (*eponzy,* Fr. *éponge*), sneakers, new clothing, and drink. Those with large sums of cash also bought zebu cattle. You could see men eyeing the herds that grazed in pastures just outside town. During this time, I also noticed a rising number of newly purchased two-handed saws for felling timber and cutting planks. Men walked down dirt roads with the saws, wrapped in leaves and twine, flopping on their shoulders. Seeing them hinted at menace to the national park.

My relationships with Mandresy, the crystal-ball maker, and his family in Mananara-ville offered other insights into the verbal *contrats* between growers and collectors in the villages and the residual colonial, as well as missionary, dynamic between Merina ("Hôva") and Betsimisaraka residents. Mandresy also tried his hand at collecting cloves from growers in 2001. Mandresy's close friend, another Merina man named Alain, headed the Mananara-Nord branch of the company and persuaded Mandresy that he could earn good money collecting in far-off, untapped villages, where farmers sold their crops cheaply. Alain had arranged for the company to loan Mandresy a total of 25 million FMG for collecting and hopefully turning a profit. From watching growers coming directly to the Ramanandraibe warehouse to weigh and sell their cloves, Mandresy observed with disapproval how the company workers sneakily weighted the scales in the company's favor and paid growers less than they were owed.

Mandresy prepared to travel with his mountain bike far from the town center and roadside villages, seeking out villages still untapped by collectors. He departed for the *ambanivôlo* (countryside) alone, not knowing how many days it would be before he and Honoré, his trusty worker, came home. Upon my return to Mananara-ville after a ten-day stay in Varary, I entered the house to find Lala bedridden and gravely ill with malaria, unable to sew, clean, or cook, and her nine-year-old daughter, Fanja, at her bedside and crying for her father to return. The situation was dire, and I was relieved I had not stayed away any longer.

Weeks went by before he returned home leaner and dusty, his muscles knotted from walking kilometers over the footpaths (having left his bike temporarily at a roadside village). Mandresy said excitedly that he had been able to purchase approximately one ton of cloves from growers in distant villages. This would make him a fortune. He was unable to carry the bags himself, so he had arranged to store all that he collected in one villager's house, a man named Augustin. He was a professed Christian, and because of this Mandresy completely trusted him.

Mandresy was one of those townsfolk who appreciated the pastoral beauty of Betsimisaraka villages and enjoyed bathing in the river and eating rice and greens on mats on the floor. And he was inspired by evangelizing in the countryside. Clove collecting was for Mandresy a means of earning monetary and spiritual profit; he rejoiced in the chance to teach

Protestantism at the peripheries of the prefecture. Plenty of Mananara-Nord villagers identified as Christian, but they tended to be more populous in Mananara-ville and roadside villages. The remote villages, accessible only on footpaths, in general seemed more faithful to ancestral ways (*fombandrazaña*), and some, such as Varary, rejected Christianity outright.

Mandresy recounted how he had prayed with his Bible over the body of an ill man, who then began to recover. He said that the villagers thought he was a healer or diviner (*mpisikidy*), which he found hugely amusing but also flattering, I think. His efforts to convert villagers worried me at the time, because in trying to do so he tacitly denigrated *fombandrazaña*. And as I was learning in Varary, villagers felt a deep aversion to things Hôva, including Christianity, but were circumspect about showing their true feelings.

After ten days or so at home, Mandresy recuperated from his trek. He took Honoré and another man to accompany him back to the countryside to retrieve and carry the clove sacks to Ramanandraibe. He was elated at the prospect of earning millions of francs, the deferred pay-off, since he had purchased the crops far below the fixed price of Ramanandraibe. The cash would completely turn around the household's fortune, which had hit very hard times. Mandresy's crystal-processing business had slacked off abruptly with the onset of the clove season, when crystal hunters were busy harvesting their trees and when many of the crystal traders, who regularly brought their quartz to Mandresy for grinding and polishing, were now collecting crops. Not able to afford an assistant, Lala sewed as much as she could between her household chores. The couple accepted my help only to a limit, which they refused to go beyond.

Again weeks passed before Mandresy and the two men returned home. When they did, Mandresy looked haggard and distraught. They brought with them another man, Augustin, the Christian villager who had promised to store Mandresy's cloves, which he had painstakingly collected from far and wide, in his own home. Augustin had in the meantime sold the lot of cloves to another collector in order to pay off some outstanding debts. The cloves could not be retrieved, nor could Mandresy's money. Augustin had confessed to the theft, remorseful and afraid of

Mandresy's special relationship to Christ. He sought redemption by accompanying Mandresy into town and turning himself into the police. Mandresy chose not to have Augustin jailed—what good would it serve?—*and* he was allowed to return to his village after confessing his crime. But Mandresy now owed Ramanandraibe 25 million FMG (over US$4,000), in addition to other debts he had racked up with others. The company director, based in Antananarivo, threatened to jail Mandresy for lack of payment. Mandresy was in a state of incredulous despair. An arrangement for repayment was finally negotiated between Mandresy and the director of Ramanandraibe, who found no better solution than allowing Mandresy to pay off the debt in installments or otherwise do time in jail. It was a struggle for the family for years afterward to pay these installments.

In these contracts the bonds of trust between producer and collector were unstable and ephemeral. The tenuousness of trust was further weakened by the opaque location of authority and its lack of commitment to consistency, which called everything into question and cultivated an attitude of catch-as-catch-can. Mandresy appeared to be an indirect victim of the culpability of his people, the Hôva (as they were called) who had colonized and terrorized the *côtiers*. While Augustin might have done the same to any other collector, in general it was felt that wrongs against a Hôva were justified by history. Ethnic tensions played out in numerous ways in Mananara-Nord, typically without direct confrontation. However, the coincidence of the spike in crop prices and the approaching presidential election, that would pit a Hôva against Ratsiraka, a *côtier*, intensified old animosities.

The looming presidential election in December 2001, the "black gold" rush that drew in outsiders, the presence of the biosphere reserve, and the propaganda being drummed up by local AREMA politicians against conservation as a way to position themselves against *vazaha* interests made peasant growers feel that the forces of dispossession were stronger than ever. On "salvage frontiers" the colonial past is selectively extracted and altered to serve the interests of power. But such selective extraction creates a disjointed and incoherent narrative about the nature of the state, whom it represented, whether it was legitimately of "the people" or complicit with outsiders.

MANIPULATING THE SYMBOLS OF STATE

Both cash cropping and conservation rely on contracts with peasants that entail a deferred realization of value. Both forms of production were girded by powerful actors tied to the state but distinct from it. In seeking to solidify trust through contracts, conservation and cash-crop exchange established states of deferral, deferred payoffs (conservation demanding a longer time span of deferred gain), which opened up possibilities for trickery and time for brewing mistrust. Because conservation and cash cropping were originally colonial pursuits, they both aggravated historical resentments and feelings of mistrust. While cash cropping could ostensibly complement conservation and development pursuits because it relieved pressure on the rain forest while peasant producers were busy processing and selling crops, the negative feelings that were intensified in the confidence games of cash cropping were transposed, often deliberately by local state representatives, into conservation activities. It was all of a mix, all shaping Betsimisaraka peasants' subjective experience of capitalism.

A episode in Mananara-ville concerning a memorial of the 1947 Malagasy uprising against the French illustrates how Western interventions into Madagascar's economy have inspired Malagasy citizens' and public officials' desire for autonomy from *vazaha* (complicated by ethnic rivalries and class inequalities that make citizens distrust officials), yet keep Madagascar's economy in thrall to the dictates of foreign donors.

Boniface Zakahelo, the *député* of Mananara-Nord and member of Ratsiraka's party, AREMA, had apparently earned (at least that is what people surmised) a sizable profit from the clove trade and wanted to show off his money. He decided to have a monument erected in the center of town, in front of two Chinese-owned stores, in early 2000 to celebrate the new millennium. He called it the Place de l'Indépendence, and, alternatively, the Place de Martyrs. It commemorated the bloody rebellion of 1947, in which Betsimisaraka nationalists massacred about 500 Frenchmen. The French had retaliated by executing nearly 100,000 Malagasy rebels or suspects, and they tortured many more. It was the bloodiest rebellion within the nineteenth-century European empires, and was one of the motivating factors of France's departure from Madagascar (see Tronchon 1986; Cole 2001). The monument was meant to mourn the tragedy yet represent the fierce spirit of resistance by *côtiers* to colonial oppression. Although Boniface's monument had nothing explicitly to do with the Biosphere Reserve, events unfolded which drew past and current events into the monument's whorl of signification.

The construction of the Place de l'Indépendence used expensive materials: a floor of concrete overlaid with marble and stone, a concrete stele flanked by two small iron cannons and topped by a bronze eagle, wings outstretched. On the stele was painted a colonial-era white pith helmet being split on either side by axes held by brown hands. Red paint spurted from the points of impact.

The monument stirred memories of and reactions to Betsimisaraka people's double subjugation, first by the Merina and then the French. It also conveyed Boniface's personal antipathy toward *vazaha.* Boniface was a young, tall, muscular man with a volatile personality that intimidated residents in town and in the countryside. Like the majority of Betsimisaraka residents, he scorned Europeans as well as Merina, particularly those who settled in Mananara-Nord. For Betsimisaraka, the older history of Merina oppression enveloped the deeds of French *colons.*

It was rumored that AREMA officials in the capital, Antananarivo, heard about and disapproved of the bloody image on the monument. The rumor was substantiated in November 2001 during a spate of "inaugurations" of new buildings in Mananara-ville intended to boost Ratsiraka's standing in the region for the upcoming presidential election. The minister of the environment, a Merina man named Alphonse Randrianambinina, was on his way to Mananara-Nord on a campaign tour, and the problem of Boniface's monument, and the bad message it sent to foreign donors, had to be addressed.

Days before his arrival, Boniface ordered the red blood in the image to be removed and the brown hands repainted white. Now the hands holding the axe and smashing the pith helmet were white. The image of violent insurgency against colonial domination was transformed into an image of brutal victimization by whites. But on whom? Malagasy people wearing colonial pith helmets? Merina functionaries perhaps? The symbolism had become opaque, nonsensical, and unstable. It also foreshadowed the rhetoric and tactics of the presidential campaign.

Boniface had bridled at being forced to alter the monument for the sake of Hôva officials and their *vazaha* donors interested in conservation. While the minister of the environment was still in town, Boniface decided to assert his autonomy as a Betsimisaraka man against Hôva-European subjugation. Not able to contain himself, he took a shotgun into the rain forest and had a videographer accompany him while he hunted lemurs, icons of Madagascar's natural heritage and national identity, as well as a red-listed species. The videotape was broadcast on Mananara television and showed Boniface proudly holding up each of his four kills. The event infuriated Raleva, the conservation agent, who thought the deputy was *mentale* (mentally ill) and vowed to lodge a formal complaint against him. Although Raleva's role as a biosphere conservation agent should have given him some pull with state authorities—after all, the crime of killing lemurs was intentionally disrespectful of the minister of the environment—nothing ever came of the incident. Boniface was left alone.

THE STATE'S EXPLOITATION OF CASH-CROP WEALTH

In Mananara-Nord, residents avidly pursued quick wealth through participation in the international cash-crop trade, which involved transactions that brought to the fore ideas about ethnic difference. The transactions also revealed the fractured nature of the Malagasy state, its dependent and resistant relationship to foreign donors, and the tendency of local state representatives to strategically mask their dependence on foreigners by stirring anti-colonial sentiments among the population.

During the period of the clove harvest in late 2001, the usual energy animated the prefecture. *Mpiavy* (in-migrants) sold their foam mattresses, wood furniture, boom boxes, and clothes in town, people engaged in drunken brawls at the local night club, thieves sneaked into the gardens of rural villagers to steal their vanilla vines heavy with pods, DJs with portable sound systems run on generators traveled around the countryside offering *kermesses,* all-night dance parties. December 2001 was also the furious height of the presidential campaign. Ratsiraka was predicted to lose, and his representatives in Mananara-ville were using all dishonest means at their disposal to ensure this did not happen.

In Varary village, every week a campaign troupe of one or another of the five candidates would visit. The villagers gathered in the center of town, sitting on the verandas of wooden houses, the children sitting on the ground outside, as electioneers stood on a granite rock and extolled the virtues of their respective candidates. The reference to cash crops in campaign speeches and the elite status of officials and crop exporters in Madagascar reflect the mutual production of capital and the state, a process into which international conservation and development institutions inserted themselves in the late 1980s.

Sophie, the wife of Sylvestre (the biosphere conservation agent), laughed hard one evening as she remembered what she had heard about the rules for the presidential election: "Those who may not vote," she said, are "dead people, jailed people, crazy people." She again giggled at the vision of dead people (*ôlo maty*) staggering to the polls. "I heard a man say that the president should stay in power for fifty years," Sylvestre declared a bit later, as we sat on the floor beside their hearth. "Why fifty years?" I asked. "Because it will take that long for the people to realize what they have endured."

Taking credit for Mananara-Nord's rising prosperity, AREMA, President Ratsiraka's party, adopted the campaign slogan, "vôkatra, vôkatra, vôkatra!" (harvests, harvests, harvests!), referring to the lucrative harvests of cloves, vanilla, and coffee enjoyed by cultivators, collectors, and exporters. Locals worried over whether Marc Ravalomanana, the Hôva billionaire vying for the presidency, would oust Didier Ratsiraka, who no matter his grave shortcomings these last twenty-odd years was at least a known quantity and was Betsimisaraka. Moral injunctions about Hôva wrongs—usually veiled from outsiders—found voice, not only around kitchen hearths and behind closed doors, but also in open courtyards and on the radio, as though expectorated by the scent of cloves.

The slogan "Tiako Madagasikara" (I Love Madagascar), the name of Ravalomanana's party, was pasted all over the taxi bumpers of Antananarivo: It read, "Aza matahotra . . . Mino fotsiny ihany" (Do not fear . . . Only believe), a biblical passage from Mark 5:36 (playing on Ravalomanana's first name). His party's name played on the name of his dairy company, Tiko, which Ratsiraka closed down in November 2001 to obstruct his rival's campaign. Mandresy and Lala's house had become a headquarters for the Ravalomanana campaign in Mananara-ville, and local supporters convened there to strategize and to reports threats made by AREMA loyalists in villages. The household was feeling increasingly scared of the vandalism (or worse) allegedly being organized by the local AREMA officials in town—the *conseil provincial,* named Eloi, and his *député*, Boniface. Knowing that if Ratsiraka was voted out of office, they would also lose their positions, they drove around town telling residents that if they did not vote "red," the color of AREMA, they would see their houses burn. They also warned all outsiders, particularly Merina, to "pack their bags" once Ratsiraka won the election, since Merina residents were nearly unanimously supporters of Ravalomanana.

Marc Ravalomanana not only was popular among the Merina of Antananarivo and Christians of Mananara-Nord but also, surprisingly, had some supporters in politically conservative villages such as Varary, who although they despised Merina people as a rule had long become disillusioned with Ratsiraka and felt Ravalomanana had a chance of winning and maybe changing the status quo. But former president Albert Zafy was much more popular. Of *côtier* origin, Zafy had unseated Ratsiraka in 1993, but was then pushed out of office by Ratsiraka's accomplices in 1996. Sylvestre, the conservation agent, claimed that he was not a strong Zafy supporter because life had been even harder during his term. "Things were even worse. Rice was more expensive. Madagascar was even poorer." He then exclaimed, contradictorily since Zafy had more regional support than Ratsiraka, that things were better now, thanks to Ratsiraka, with revenues from cloves and vanilla.

Albert Zafy's campaigners came to speak in Varary weeks before the election. Their rhetoric addressed the problem of alienation in the countryside deriving from the dependency on the cash-crop market and land expropriation by outsiders:

> Regarding crops, in the past coffee was 10,000 FMG per kilo because Zafy Albert was in power. But now it's 500 FMG per kilo. And the very high price of cloves? It's not Ratsiraka who has raised the price but, again, it was Zafy Albert's idea. We in the countryside don't have sources of income apart from dance parties [*baliña*, from the French, *balle*]. Plus we have work to do concerning the school and the road because we should be paying someone to do all the work of the *fokonôlona* [an administrative collectivity of villages]. But because we don't have money, we're obliged to do all the work ourselves. So, if we vote for Zafy Albert we'll earn forty million Malagasy francs for ourselves. We won't need to work the *fokonôlona* so hard anymore; instead we can take the forty million to pay someone.
>
> Next is what we have right now. And the rich come like that, they come before the poor. They, the rich, without speaking to the poor, they come to measure the land of the poor, and once they're finished measuring

> the land they leave. And then the rich return and go to the poor and say, "leave, because this land where you are now is ours." The rich are angry and they say this about our land where we are. And the poor go to lodge a complaint but when they get there the papers of the rich have already been left with the state. And that's how the affair is finished. The land of the poor is taken by the rich. (Field notes 12/01/2001)

Campaigners competed over who was the most populist president. *Mpiavy* (in-migrants into Mananara-Nord from surrounding areas) and foreigners were lumped together with the land thieves. Presidents appeared to be the deciders of global market prices. The upcoming election would give voice to people's value of landownership and group solidarity against expropriation and colonialism in all its guises.

In the week leading up to the election, campaigners of each party made the rounds through the biosphere reserve wearing T-shirts with their candidate's logo and promising a future of growing prosperity. AREMA campaigners attributed the high prices of cloves and vanilla to Ratsiraka's policies, beseeching the audience of Varary village: "How many sheets of metal [*taoliñy*] now deck the houses? Thirty thousand [sheets of] *taoliñy* have entered Mananara this year! How many bicycles do you now see on the road between Sandrakatsy and Mananara-town? How many foam mattresses [*eponzy*] now soften the beds?"

Varary residents gathered in the central courtyard of the village to listen to campaigners make their pitches on top of a large flat rock and participated in open discussion after each speech. The eloquent AREMA campaigner spoke:

> We already know the mentality of Hôva. I say to you well that while we were studying at the university, there were Merina that studied with us, saying "you blacks, you are slaves of us Merina," they said. That was what the students did to us. And I can't ever forget it, what they did to us at that time. So what we have since we've been on our own is to vote, because it's hard to give up what one gets used to. And also, take a good look, if one wants to switch mates all the time, one truly can't accumulate anything when constantly changing spouses. (Field notes 12/09/2001)

Sylvestre was the first in the audience to respond. He proceeded to defensively praise Ratsiraka:

> it has already been a long time since cloves here have produced a lot, but never was the price like it is now. With prices like this here, it's rare to see houses without foam mattresses. Lots of people have Sonys and batteries.

> And I am not putting down Varary but it was rare to see people who could ride a bicycle. But now there are already a lot of people who know how to ride a bicycle, and also a lot who have bicycles. And there's no one else who brought about all this of which I speak, no one but Ratsiraka. Ratsiraka raised the price of cloves, but no one else! (Field notes 12/09/2001)

Following Sylvestre's discussion, Paul, the brother of Jafa, the other conservation agent, chimed in. He also sought to protect his interests by advocating political conservatism and voicing support of the president: "Be careful about seeing a new mate and leaving the old one. Watch out, ééééé! And think it over well. When one changes spouses all the time, one can't save. Always constant new things." The rhetoric of fearful difference referred not only to ethnic identity (beware the Hôva) but also to the future itself. By comparison of the head of state to a spouse, the state became intimate, drawn into the sphere of domestic life. It was taboo to marry Hôva; to do so invited danger. Paul's metaphor alluded to the danger of voting for Marc Ravalomanana, a Protestant Hôva. "Beware the slave masters," campaigners said of the Hôva candidate. "The Hôva have always been liars [*mpandika*]."

Social change in Mananara-Nord was in the early 2000s bounded by the tension of the island's clove corridor being rapidly expanded and hooked into international markets, while the national power struggle over the presidency in late 2001 closed the island off to tourism for several months. Ratsiraka, refusing to vacate the presidency and attempting to squelch the massive crowds in the capital celebrating Ravalomanana's victory, had plotted to blockade the island's roads and stop domestic and international flights. For a time, the roadblocks and flight suspension benefited the Chinese shipping companies with routes along the east coast, who could outcompete drivers with their monopoly and the transport of crops, as well as other goods, to and from the port of Toamasina.

CONCLUSION

Conservation rules were contracting the horizon of the future, was the commonly held belief. "Better dead tomorrow than dead today" (*aleo maty amaraiñy toy izay maty iníany*) was a proverb some peasants uttered when asked why they did not support conservation. Legally prevented from clearing forest, people were forced to replant the same plots repeatedly, and rice harvests were diminishing due to eroded soils. This deepened poverty, preventing accumulation that would enable people to buy, among other goods, zebu cattle, which are essential for laboring the soil for farming, as well as for offering sacrifices to the ancestors. Dead ancestors ensured their descendants' future well-being. So in these respects, global conservation, represented by the biosphere reserve, was intruding on the sensibilities of residents as time-space compression.

Conservation efforts in Mananara-Nord were dependent on forms of production that brought quicker and more reliable returns than the prospect of species proliferation or the trickling in of revenues from ecotourism. The slower pace of conservation and development was hooked into the faster, cyclical temporality of cash cropping. The cultivation, processing, and exchange of cloves and vanilla, in particular, comprised an important source of cash for peasants of Mananara-Nord, placing them along a commodity chain that had valorized the space of rain forest since the late nineteenth century. Aromatic trees and vines had far greater value to Betsimisaraka residents than did the slow-growing endemic species saplings cultivated and sold cheaply in the two nurseries established by the ICDP.

Cloves and vanilla in the early 2000s brought eastern Madagascar into the compressed space-time of globalization. Residents of Mananara-Nord perceived a rapidly changing landscape and expressed misgivings about their future livelihoods, as migrants rushed in, disrespecting local ancestral customs, sometimes thieving, and creating an atmosphere of instability, risk, and recklessness. Despite the fact that cash cropping did not appear to impinge on the integrity of the rain forest, the culture of cash cropping fortified people's sense of defiance against outsiders. This feeling was manipulated by AREMA officials trying to keep their hold on power. Because crops brought rapid wealth, politicians used them in the rhetoric of Otherness that defined the history of Betsimisaraka-Merina and European-Betsimisaraka relations. Crop wealth heated the politics of ethnicity, stoking the embers of antagonism among members of groups that comprised a colonial typology. Otherness was invoked in the words *Hôva* and *vazaha,* which worked to focus people's imagination, to deflect people's meditation on the exploitation of Betsimisaraka by Betsimisaraka.

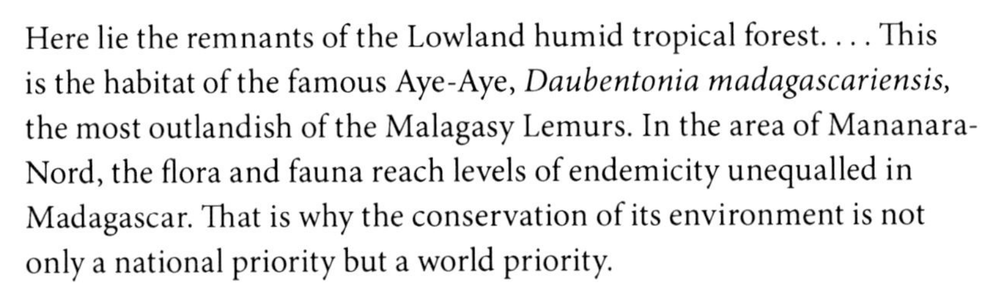

Here lie the remnants of the Lowland humid tropical forest. . . . This is the habitat of the famous Aye-Aye, *Daubentonia madagascariensis,* the most outlandish of the Malagasy Lemurs. In the area of Mananara-Nord, the flora and fauna reach levels of endemicity unequalled in Madagascar. That is why the conservation of its environment is not only a national priority but a world priority.

—UNESCO's brochure for the Mananara-Nord Biosphere Reserve

6 How the Dead Matter

The Production of Heritage

In Madagascar, several cultural and natural heritage sites have been included on UNESCO's World Heritage List, including the Royal Hill of Ambohimanga, consisting of a royal city and burial site, the cathedral-like limestone formations (*tsingy*) of Bemaraha, and the rain forests of Atsinanana, "relict" forests of the east coast. World Heritage sites possess at least one of ten criteria of value, including such things as exceptional biodiversity, ecological service, beauty, historical and archaeological significance, and creativity (UNESCO, World Heritage Convention, http://whc.unesco.org/en/criteria). UNESCO's World Heritage Convention webpage defines "heritage" as "our legacy from the past, what we live with today, and what we pass on to future generations. Our cultural and natural heritage are both irreplaceable sources of life and inspiration" (UNESCO World Heritage Convention, http://whc.unesco.org/en/about/). The World Heritage preservation program has identified between eight hundred and nine hundred built and natural sites, as new sites are being evaluated for possible inclusion on the list (Breidenbach and Nyíri 2007:322). The geographical sites and material culture that constitute world heritage are further disaggregated by indigenous cultural formations and endemic species, which denote "cultural heritage" and "natural heritage" respectively (Brown 2004:49). By reifying discrete elements of a place, heritage preservation particularizes the homogenizing force of globalization in different geographical locations and brings into metropolitan consumers' view digital data images and texts about life at the global peripheries.

As John Collins (2011) argues, the heritage rubric essentializes nation-states because the intellectual and physical labor of turning places into heritage sites produces national identity and thereby transfers control over communal lands or neighborhoods from local communities to the state (Collins 2011:125). National heritages, governed by states, are also elements of a world heritage, their value universalized and subject to the juridical authority of global institutions. Regarding Madagascar's national natural heritage, for example, Noël Randrianandianina, the director of ANGAP, expressed in an interview with *Madagascar Magazine* in 1999 that Madagascar's protected areas are at once "a source of national pride for present and future generations" and a benefit to "all of humanity" due to the high degrees of biodiversity and species endemism (Raony-Le Boubennec 1999:15). Ironically, the prospect of a lucrative tourism industry in Madagascar based on the exchange of its natural heritage appealed to the Malagasy state, but also had its downside. A North American USAID employee in Antananarivo told me in February 2001 that officials feared the "corruption of culture" that came with the growth in international tourism in Madagascar, alluding to the social effects of sex tourism.[1]

With the "inscription" of a site on UNESCO's world heritage list, the universal value of certain vestiges of the past manifests as a product of intellectual conservation labor. Collins argues that the distillation of historical processes into essentialized identities, defined as heritage, serves to reseed the ground with the relational dynamics that got us here (at a place where we must salvage heritage) to begin with (Collins 2008:288–289). Heritage preservation is part of an integrated set of initiatives to "valorize value," as Marx (1977:991) put it, by resignifying the products of historical labor processes and "entic[ing] consumers to participate in the resolution of capitalism's environmental contradictions through advocacy, charitable giving and consumption" (Igoe et al. 2010:487).

The designation of heritage sites and identification of heritage objects abstract and reify social relationships to produce apparently stabilized units of time and space. Around these, social activities proceed. The determination of the criteria of value gravitates toward that which is old and rare. Old age and rarity represent "purer" forms of culture and biology. The degree of an object's purity derives from its apparent proximity to an original source, making it more authentic. The search for authentic experience motivates international tourists, especially those who visit poor countries and biodiversity hot spots. So strong is the desire to behold landscapes that seem relatively unaltered by human activity, and to interact with cultural groups that seem relatively uncorrupted by the trappings of modernity, that many tourists are willing to suspend disbelief that preserved and restored sites, or reenacted traditions, are glimpses into a more authentic time (Frow 1997).

But what happens when "authentic" cultural heritage does not accord with the criteria for universally valuable nature/culture identified by UNESCO? The practices that are valued by cultural groups as distinctive markers of their group identity frequently conflict with these criteria because they jeopardize the natural heritage of the nation and, therefore, the world. What happens when the labor processes of diverse groups of people, living relatively close to "nature," impact the land and species, and imbue these outcomes with particular meanings that diverge from the value hierarchies of outsiders?

In Mananara-Nord, the conservation of natural heritage—endemic species and primary habitats—was not necessarily the types of heritage that Betsimisaraka peasants found to be important or worth bequeathing to their descendants. On the contrary, the establishment of the biosphere reserve as a form of world heritage preservation met with resistance because it imposed constraints, rules, and unfamiliar ideas that threatened northern Betsimisaraka people's sense of heritage, a concept that approximates that of "ancestral custom" (*fombandrazaña*).

Respect for and obedience to "ancestral custom" entailed ritualized work and ritual practices that are central to all Malagasy people's identity as individuals and imagined communities. In the early 2000s, Betsimisaraka villagers of the biosphere reserve created their cultural and natural heritage by honoring *fombandrazaña,* but in so doing they jeopardized the longevity of endemic species and undermined the restoration of ecosystems based on approximations of what they looked like before human settlement.

Conservation labor often conflicted with customary practices; while these commemorated a social history and constituted the cultural heritage, some of them also depleted biodiversity. Villagers of the biosphere reserve saw their ritualized interactions with trees, soils, waters, and animals as reenactments of their ancestors' practices, altered by necessity sometimes in substance but not in form. Yet they also perceived these customs to be succumbing to accelerating change as outsiders moved into the area, expanding the size of Mananara-ville, as well as other roadside villages of the prefecture. The influx of outsiders and developments in town increasingly tempted young people to abandon ancestral ways, much to the anxiety of village elders.

Conservation planners worked to suppress the reproduction of *fombandrazaña* when they tried to intervene in the practice of *tavy*, the dominant mode of production and reproduction in the eastern forests. Through *tavy*, Betsimisaraka cultivators (*mpambôly*) valorized land, forged their group identity, and ensured the succession of their genealogical line. In addition, they memorialized acts of resistance to the colonial state, which had banned *tavy* in 1896 and penalized Malagasy people for breaking the law. Betsimisaraka people's production of heritage occurs through ritualized agriculture, funerary practice, and respect for *fady* (taboos). In all of these practices, people solicited the active collaboration of dead ancestors, symbolically grafting external nature onto the family tree.

Through the lives of several conservation agents, Sylvestre, Jafa, and Ali, I describe how the operative principle of heritage manifested in Mananara-Nord. I then recount a case in which the production through selection and preservation of "world heritage" conflicted with the privileged form of heritage of a small locality, Sahasoa, which had been incorporated into the biosphere reserve as its marine reserve.

TWO FAMILIES

The establishment of the biosphere reserve and national park in 1989 caused tensions between the families of Sylvestre and Jafa in Varary village. Sylvestre, the *Parc Responsible* of Varary and the son of the *tangalamena*, Rasonina, was going to inherit his father's role upon his death. The *tangalamena* was traditionally the eldest male of an extended kin group. Due perhaps to his son's and grandson's employment by the biosphere project, Rasonina claimed to be a strong proponent of conservation and often recited to me the ecological benefits of protecting the forest. Yet despite Rasonina's support of the biosphere project, the majority of Varary inhabitants had mounted a fierce but unsuccessful opposition to the biosphere reserve. Many residents claimed to have had land subsumed within the national park, even though, as one farmer admitted, the Verezanantsoro forest was technically off-limits to farmers before the park's boundaries were delimited because it was a state forest concession, and anyone who wanted to clear a portion to cultivate needed to apply for authorization from the regional forest service station in Mananara-Nord.

Jafa's father, Lema, had never bothered concealing his disdain for the biosphere reserve. Lema claimed that his land had been stolen by the biosphere reserve people, and he was furious. My research assistant at the time, Zosy, a young Betsimisaraka woman from Mananara-ville who was living in Varary as the common-law wife of the unofficial school teacher, told me that Lema was rumored to be a sorcerer (*mpamosavy*). She also said that the same suspicions were whispered about Rasonina. It seemed that all gray-haired individuals were called witches. The tense relations between these two elders were familiar to the village. Their pro and anti stances toward the conservation effort were inversely manifested in their sons' agricultural practices.

In spite of his anti-conservationist father, Jafa practiced agriculture in a way that would make the environmental authorities in the capital proud. Jafa and his wife, Nirina, had four children between the ages of infancy and eleven years old. In addition to farming and Jafa's employment with the biosphere project, the couple ran a small store attached to their house, where they sold sundries such as matches, candles, plastic jugs, dried beans, bottles of soda, and shots of rum. Jafa's household did not own *tavy* or *jinja* (as the land for the cultivation of hill rice was called in Mananara-Nord). Instead, Jafa possessed paddy (*hôraka,* which is rice cultivated in flat marshy land or sloped, irrigated land), in which he had carved wide, flat terraces to maximize the sloped space and prevent soil erosion.

As it happens, this is precisely the approach promoted by the ICDP, so Jafa's plot was heralded by the ICDP bosses as a model for other villagers. In the *hôraka,* he usually worked alongside his wife and her sister, while his older children attended school and the women traded off keeping an eye on the babies and preparing the midday meal. Although Jafa did not discuss it, his relations with his extended family appeared somewhat distant. His father was troublesome, as I will explain later.

In Mananara-Nord, villagers' plots of *tavy* lay at various distances from the village center, sometimes hours away, which made walking a time-consuming aspect of the labor process. So no matter how far away a family's fields, they considered the village their proper home even if during certain periods of the agricultural cycle, such as sowing and harvest, families might sleep in crudely constructed shacks on their plots rather than make the trek back to the village at the end of the workday, a trek that could take hours. The yearly cycle of clearing forest and scrub, sowing, weeding, and harvesting crops, as well as cooking meals, and at times sleeping in the *tavy* plot were activities that made intimate immediacies out of nature: the thickness, shadiness, itchiness, and edibility of plants, the beauty, danger, or irritation of insects, the furtiveness, speed, peskiness, taste, and size of animals, the cleanliness and coolness of water, the ferocity of heat, and the dryness, wetness, smoothness, temperature, and length of footpaths affected people's quotidian existence.

From November through June, spanning the time of sowing and the time of the rice harvest, the workday for residents of Varary began before dawn, when the cocks began to crow and thick mist still hovered above the wild grasses. The air thudded rhythmically with the sound of women and girls pounding rice for the midday meal in their large wooden mortars.

Sylvestre owned two large parcels of *hôraka*, and two parcels of *tavy*. He had bequeathed land to his adult children as well—he and Sophie had a total of nine children ranging from twenty-seven to seven years of age—some of whose land consisted of the fringe of primary rain forest. This land, Sylvestre insisted, was not part of the biosphere reserve's national park.

Unlike the cultivation of other crops, rice cultivation requires specific rituals (*fômba*). For example, during certain days of the week, individuals were prohibited from working on their *tavy*. It was *fady* (taboo). Betsimisaraka clans have respective *fady* days regarding rice work, whereas domestic labor and cash-crop labor are exempt. In Varary, for example, the *tangalamena*'s extended family could not work in the rice fields on Tuesdays and Sundays. On these days, the family would instead tend to chores in the village, visit people in other villages, walk to the roadside town of Sandrakatsy to buy or sell goods, or work in their gardens or cash-crop groves.

Sylvestre told me that before the paddy is sown, it is customary to plant a *mahatambeloña* ("dead-living," or "to make dead-living")[2] tree, a species with quickset branches, at whose base the landowner places offerings of a cooked chicken and fermented cane juice (*betsa*). Whereas a ritual had to be performed before paddy rice was sown, the main rituals for *tavy* were performed after the harvest. Landowners (*tômpony*) were obliged to sustain good ancestral relations to and through the land. Inasmuch as the performance of commemorative rituals in the agricultural labor process was a way to ensure ample harvests and health, these rituals, and the whole complex of practices tied to *tavy*, also commemorated colonial resistance. As Gillian Feeley-Harnik (1984) discusses regarding the Sakalava of the northwest coast, the preoccupation with dead ancestors is about loyalty to one's own in the face of repressive forms of power and authority. She argues that Malagasy people's interest in the dead evolved as the French attempted to destroy native Malagasy institutions. The implementation of new forms of labor and loyalty through the language of law under the colonial state compelled Malagasy to assert their own "lingua franca of power," which valued loyalty to dead ancestors: "Interest in the dead . . . derives from common historical roots in experiences of the body and its labor that are inseparably physical, social, and philosophical, experiences that might therefore be summarized in the phrase *the political economy of death*" (Feeley-Harnik 1984:2).

In Varary, residents paid honor to this historical heritage as they carried out the cyclical tasks of agriculture by leaving sacrificial offerings of chicken, rice, and rum on stone slabs on the farm plot for the ancestors. Zebu cattle were sacrificed for deceased parents as well, but this during a larger ceremony called *tsaboraha*.

Sylvestre and his wife, Sophie, had on one of their *tavy* plots an offering site for ancestors, a wooden stake (*fisokina*) with a stone slab at its base (*lokambato*, "rock site"). Three smaller stone feet supported the slab. At harvest time, the family would present offerings of meat, rice, cane juice, and moonshine (*kavilava*) on the *lokambato*. Meals would be presented on palm-leaf plates: a spread of cooked rice topped with cooked chicken legs or pieces of beef. I was told that meat was left on the *lokambato* only temporarily, then consumed by the family. Ancestors, I was told, "can't eat. They only ingest the aroma." Descendants of an ancestor could also decorate the tip of the wooden stake with zebu horns and chicken heads to record past offerings.

One day in December 2001, my research assistant, Zosy, and I accompanied Sylvestre and his family to sow his brother Dely's plot, which lay about two hours on foot from the village. Families working together on each other's plots of land reflected the ethic of mutual obligation and solidarity captured by the term *fihavanaña* ("familial relations"), an important ethical principle throughout rural Madagascar. I describe the atmosphere of familial labor to give a sense of why Sylvestre, Jafa, and other village-based conservation agents could not easily separate themselves from the moral economy of *tavy* and perception of nature intrinsic to it, even though their ICDP jobs implicitly expected them to intellectually alienate themselves from the cosmology of the village.

As we walked with Sylvestre's family along the footpaths to Dely's plot, members of the group would gather wild edibles for the midday meal. Martine, Sylvestre's twenty-two-year-old daughter, noticed a gurgle in a mud puddle beside the trail. She scooped her palm into the bubbles and captured several *sambiremy,* a species of tasty, small fish, as well as a small tadpole, which she flung into the brush. Martine wrapped the live fish in a leaf and we continued. The trail was grown over with at times waist-high scrub containing *anakolokoloko,* a plant with small purple flowers named after the sound of a turkey's gobble, and *totroka,* a plant used for stomach trouble as well as "toilet paper." The longoza, a plant with red sour fruit, grew high, about ten feet tall. Martine was also on the lookout for cicadas (*pinjy*), whose opaque red eyes and painted bodies looked like ceramic broaches. Her three-year-old daughter Filo liked to play with cicadas, squeezing them lightly to make them buzz. When dewinged and fried, *pinjy* were also favorite snacks.

We stopped at a sharp bend in a stream that carved out a little peninsula at a tail of the burnt *tavy.* The camp was shrouded by shrubs, on which the family set up camp. Reeds mats lay on the ground and an umbrella shaded the infant of Boavy, the wife of Sylvestre's brother, as she tended to the fires under the pot of cooking rice. Zosy and I, along with

Martine and Martine's three-year-old girl, sat on the mats and watched from afar as men in the field doused tree stumps still smoldering after the burning.

Families often ate the midday meal together in the shade of their "long house" (*traño lava*), a makeshift shelter made of wood, bamboo, and dried ravinala leaves. On a day out with Sylvestre's nuclear family, everyone stopped working to convene around 11 AM for a meal of *loja*, a small groundnut, boiled in coconut milk, as well as cooked *angivy*, an intensely bitter fruit resembling a small tomatillo, and *hôrita*, the thick, bitter heart of the ravinala palm. Cooked rice and bowls of greens, squash, or other vegetables were spread upon green ravinala leaves. Boiled rice water quenched thirst.

The food was consumed with ravinala leaf "spoons" (*sôroko*). After the meal, the leaf spoons had to be ripped at their bowls to show to passersby that the meal had been consumed by mortals and that invisible ancestors were not in the process of dining. The ravinala palm and the longoza plant, which grew abundantly on fallow ground, had multiple uses as food wrappers, plates, "tablecloths," and cups. The expression, "a longoza dies, a replacement grows" (*longoza maty sintoño mandimby*) refers to lineal succession. Sophie, the wife of Sylvestre, explained the proverb as follows: "When grandfather dies, grandfather is replaced by father. When father dies, he is replaced by son," and so on (Field notes March 15, 2001). Eating meals together confirmed for Betsimisaraka the bonds of family and the purpose of labor.

During harvest time, people with large plots often sought extra labor. At the end of a workday of harvesting, the landowners customarily expressed their thanks to helpers by inviting them to share in the evening meal and, especially, in drinking of *betsa* (fermented cane rum) at their home in the village. After eating and relaxing for a spell, the family grabbed their sowing sticks, sharpened the points with their machetes, and returned to the field.

I noticed one day, while eating in the long house erected next to one of Sylvestre's rice paddies, that a piece of wood in the structure was etched and painted with the words PONT MAHAVOTRA. A land marker of the biosphere reserve, a bridge that no longer existed, was thus incorporated into Sylvestre's "other" life as rice farmer. It was an apt symbol for his position, as well as the position of all residents of the biosphere reserve who lived off the land and were encouraged to see it in a new light, as rich with unique species and biogeographical formations, yet ephemeral unless people adapted the *fômba*.

ACTING AGAINST ANCESTORS

Jafa's father, Lema, owned *fanómo*, the word for land that lies next to tombs. This is very powerful land, and those wishing to sow rice on it must perform a ritual before clearing the vegetation and burning it. *Fanómo* was identified by stone boundary markers instead of tree rows. Families wishing to burn vegetation on *fanómo* had to perform the *fangatahaña lalaña*, "request to pass through" from the ancestors, and *ravaraty*, "the ravaging of dead vegetation." During the *ravaraty*, people would beg permission to burn from the ancestors as well as the *tsiñy*, ancient spirits or "forest divinities" (Fanony 1975:106). *Tsiñy* (an evolution of the word *jiñy*, I assume) haunted the forests, waterfalls, deep pools, imposing trees, and roads, and could treat the living with malice, kindness, or indifference. While people regularly commemorated their ancestors, they directed rituals to *tsiñy* on only rare occasions.

Another resident of Varary, a middle-aged man who served as the administrator of the hamlet, explained that *tsiñy* were "ancestors who died a very long time ago." They were literally imagined as ancestors gone wild. No longer tended to by their living descendants, they were also no longer bound to the bonds of reciprocity between the dead and the living. People nevertheless wanted to appease *tsiñy* for their self-protection. In other villages, people might offer honey to *tsiñy* before burning forest, uttering: "Here is some honey so that you give us your blessing. Leave these environs if you do not want to be burned" (Fanony 1975:106).

An elderly *tangalamena* of a village a few miles from Varary told me in September 2001 some more about how one proceeds to farm *fanómo*. When one first steps foot on *fanómo*, it's necessary to lay some honey on the offering stone of the *tody finomaña*. "And even after you get rice," he continued, "there is still another custom, killing a chicken and leaving it on the offering stone of the *tody finomaña*." The name of the ritual, *tody finomaña* ("punishment by drinking the poison of the tangena nut"), derives from the historical practice of "trial by ordeal." People would administer the poisonous nut of the tangena shrub (*Tanghinia venenifera Madagascariensis*) to determine whether or not a person suspected of being a sorcerer was indeed one.[3] *Finomaña* signifies the action of drinking the poisonous juice of the tangena nut (Abinal and Malzac 2000:281). The nut is also known as the "Madagascar ordeal bean" (Felter and Lloyd 2004).

Apparently Jafa's father, Lema, had neglected to sacrifice a zebu before burning his own *fanómo* despite owning an ample herd of zebu cattle. Instead, people gossiped, he fulfilled his ancestral obligation by substituting a person for a zebu. Not just once but twice, and these sacrifices were his own grandchildren! Even Jafa believed his father to be a sorcerer for having done such a thing, since his wife miscarried twice.

Lema died unexpectedly in December 2001 while drinking *mafana* ("heat") with a crowd in another village. *Mafana* is the name of local moonshine, made in large wooden stills, which has an extremely high alcohol content. His family along with other Varary residents suspected he was poisoned deliberately by someone.

People claimed that at Lema's funeral his wife appeared noticeably calm, exercising restraint in her grief, which made people believe she was not that sad to see him gone. It was customary at funerals in the region for women to cry loudly. Lema's children likewise seemed less than distraught. Yet once dead, despite his possible dabbling in witchcraft, Lema became an ancestor like any other and had to be treated with the respect and care due to all ancestors. Tending to the dead was an important way in which people rooted themselves to territory, so cultural heritage, such as mortuary practice, was inextricable from natural heritage, including the land and the natural features and species on it that were incorporated into ritual practices.

ROOTING GENEALOGY TO THE LAND

Oral histories told by Varary residents chart social and cultural adaptations to an evolving landscape. On the matter of mortuary practice, people's use of large hollowed-out trees for coffins succumbed to the reality of the depletion of old-growth trees. According to the forestry official Jeannelle's 1900 report, coffins would be alternatively buried, raised, or left exposed. Some coffins "are buried, others are placed on the ground and finally there are those that rest on supports; the first are the coffins of people who while they were alive did not dread either obscurity or humidity; the second are those that did not fear the gloom; and finally the third are those who were horrified by the one and the other."[4]

Funerary rituals facilitate the transformation of the living into ancestors, a preferable state to ghosts (*angatra*), who cause terror in the living. Communicating and paying tribute to ancestors through ritual practice at sacred sites, and through symbolic associations of land and kin, reproduce *fombandrazaña*. They ensure cultural survival.

Tsimifira, a middle-aged man who also resided in Varary, described mortuary practices in the region that were in the process of becoming obsolete because of the changing terrain and custom. His descriptions revealed the symbolism of genealogical rooting to territory, as well as the way environmental degradation transformed practices that were species-specific, such as the construction of coffins made from old-growth trees.

Tsimifira recounted that it had long been the case that when an adult person died, it was customary to fashion a coffin from the trunk of one of three kinds of tree (*kakazo*): *ambôra* (*Tambourissa* sp.), *vilompony* (various species of *Dalbergia*, or rosewood), or *hazovola* (*Dalbergia humberti;* "money wood"; also known as palisander). These hardwoods decompose slowly and thus protect the corpse from inclement weather and animal scavengers. According to Mangalaza (1994), the coffins made from the *ambôra* tree, once abundant but now scarce (and scarce at the time Mangalaza wrote), traditionally served as a temporal measure to determine degrees of relatedness among village members and to avoid incestuous marriages through exogamy. According to customary regulations, collaterals of the male ancestral line may marry after the passage of nine generations to avoid incest, while six generations must pass for the uterine line. *Ambôra* wood is thought to rot (*lô ambôra*) in nine generations' time. Fulgence Fanony (1975:183) also writes that if after unearthing the tomb of a common ancestor, the casket had rotted enough to reveal bones, the marriage was acceptable (Fanony 1975:183; Mangalaza 1994:38).

Tsimifara, who was in his mid-forties, told me that he remembered when the trees had been so large that people only used the center of the trunk for a coffin wood (*hazo*), lopping off the treetop and placing an offering of wild honey and rum (*toaka*) on the stump. A body-length hole was carved out of the trunk's midsection, and the *hazo* was buried underground.

To mark the grave, two X-shaped wooden stands, forked branches of *kesikesiky* or *hasimbola,* types of trees, were planted at each end of a buried coffin, cradling a cross-pole. The structure was called a "protector from cattle" (*fiaroaomby*). Eventually, these branches take root and sprout—perhaps symbolizing lineal flourishing. The cuttings of the employed tree species, *kesikesiky, hasiña,* or *hasimbola,* are the same ones used as boundary markers around individuals' landholdings. Feeley-Harnik notes that among the Sakalava of the west coast, shrubs called *matambelona,* or "dead-living," were planted with those called *hasina,* or "generative" (1991:169, 184), whereas in the village of Varary in the Mananara-Nord region, Betsimisaraka pronounced the shrubby tree *mahatambeloña.* The prefix "maha" in Malagasy typically denotes "that which makes," and *veloña* signifies "living" or "alive."

Burial practices varied from village to village in the biosphere reserve. In some areas, funerals had been heavily influenced by Christianity; others, as in Varary, resisted Christian practice and adhered to custom. Varary residents maintained familial tombs that were shrouded in remnant stands of primary forest. Whenever one saw tufts of standing forest in an expanse of otherwise shorn mountains, one could be certain they harbored a graveyard. It was *fady* to burn forests sheltering burial grounds. Conservation biologists are aware of this practice and have identified it, like the taboo against eating certain species of lemur for many families, as examples of indigenous conservationism (see Keller 2009).

Funerary practices for infants largely still followed custom in the early 2000s, although residents claimed that custom was giving way in certain areas to infant burial in coffins. Dead infants up to four months of age were traditionally put to rest differently than adults because they did not yet possess teeth. The lack of teeth indicates to Betsimisaraka people that a baby does not yet have "bones" (*tahôla*) and is thus not yet fully human. Infant corpses were wrapped in cloth and placed in a living tree near the familial tomb, in the leaves of a fecund orchid plant or other epiphyte at the fork of two thick branches. The corpse would be gradually eaten and transported by pieces into the air by birds, the cloth getting entangled in the branches of the canopy. The epiphyte or orchid plant served to stabilize the corpse as well as the leaf cup placed beside the corpse and filled with the white, gluey sap that flows in the stems of certain plants. The white sap represents a substitute for mother's milk. As Rasonina explained in January 2001, "milk" is offered because the baby must eat every day. One utters to the infant corpse when offering the cup, "that's your milk. Don't disturb your mother at night in dreams. You are with your grandparents now."

Family members asked their ancestors to care for the infant. Stillborn babies, or babies who died on the day of birth, were instead placed at the base of a tree, with a cup of white sap, and encircled with a small wooden fence made of living cuttings of species that did not produce the "milk" sap, perhaps to enable the infant's weaning. Many people in Varary and neighboring villages believed that burying a baby in a coffin would obstruct the mother from future childbearing.

Unlike offerings left for ancestors in the rice plots, which consist of cooked rice and meat placed on *ravinala* or *longoza* leaves, offerings submitted in the forest-shrouded graveyards consist of manufactured, durable products such as cloth, enamel cups and plates, and coins. Family members of a deceased person leave offerings of rum (*toaka*), clothing (*lamba*), and coins on stone slabs at the entrance of cave tombs or inside *lava-bato* (dug holes sealed with stones). If the deceased was someone of great distinction, however, a "zebu head" (*lôhanaomby*), actually the horns, would be put near the burial site as "proof" (*profo*) of the person's high value. Outside the shady space of burial grounds, offerings to ancestors mostly consist of more ephemeral foodstuff: rice and meat (zebu or chicken), in addition to rum and fermented cane juice.

My observations of practices in the biosphere reserve revealed that the ritual work of funerals, rice farming, and celebratory events such as the *tsaboraha* bonded rural Betsimisaraka communities to the land and its creatures. At multiple sites, earthy symbolism interwove human and vegetative processes. Mediating the "seen and unseen worlds" of the dead and the living (Feeley-Harnik 2004:45), plants, trees, animals, and stones generated human descendants. As living people did things for dead ancestors—work the land, offer crops and meat, tend to tombs, respect customs and taboos—ancestors also worked for their living descendants, overseeing the fertility of wombs and rice fields. Subsistence labors reproduced Betsimisaraka life through an idiom of ancestrality—

that was an ideology of kinship that made the past a dynamic force in the present.

SACRED SPECIES OF THE BIOSPHERE'S MARINE PARK

At a much smaller scale of energy and interest, the preservation of cultural heritage was part of the conservation and development mission. For example, at the sites of other protected areas in Madagascar, such as Fenerive-Est, a large town located north of Toamasina, ICDPs were instrumental in creating small "ethnological museums" either in town (as in Fenerive-Est) or in off-road villages near national parks. During my stay, Mananara-Nord did not have an ethnological museum, but the idea was bandied about now and then. These small museums were typically stocked with traditional agricultural tools (such as dibble sticks, spades, and grain baskets), seeds, stones and beads used by diviners, wooden looms used a century ago (and still to some extent in remote villages) for weaving raffia palm fibers, and other artifacts of Betsimisaraka village life. Museums have long served to encase bits and pieces of material culture abstracted out of the current of history. The ethnological museums were designed to enhance tourists' visits to protected areas by familiarizing them with selective aspects of rural Madagascar. However, these were negligible sites, easy to miss, and possessing little touristic allure compared to the nature reserves.

In Mananara-Nord, biosphere administrators thought "village stays" a more plausible thing to offer tourists who made the trek into the national park. The idea was that tourists would stay in the *gîte* in Varary, would eat a meal with Sylvestre or Jafa and their families, and would get a taste of real life in the village. This happened only a handful of times while I was staying in Varary; for the most part, the condition of the *gîte* and the material realities of rural poverty left European and North American tourists pining for beachside bungalows and pleasant tourist-grade hotels. The villages were part of the biosphere reserve, but subordinate in value to the biodiversity and aesthetic beauty of the rain forest. In the biosphere's marine reserve, however, tourists were introduced to an interesting experience of heritage preservation that called into question definitions of heritage, authenticity, and the moral hierarchies of value imputed to the nonhuman world.

Over the past two decades or so, foreign journalists and scientists in Madagascar have remarked on the trend of Malagasy people transgressing or abandoning certain animal proscriptions that have long served indirectly to protect species from endangerment or extinction. Individuals or households who may have abided by *fady* against eating sea turtles or lemurs, for example, have been giving way to the realities of species depletion and the need for new sources of protein (Colding and Folke 2001; Jones et al. 2008). As concerns animal taboos in Madagascar, an experience of Ali, the conservation agent in charge of the marine reserve on the east side of the biosphere reserve, brings to light a conundrum of heritage-making in biodiversity hot spots.

The story of Ali protecting the islets of the marine reserve relates how residents of the region abided by animal taboos that they considered of utmost importance because they protected the moral order of their universe. However, people's obedience to the animal taboos of the marine reserve undermined the work of restoring the landscape to an imagined pristine state, as well as the work of protecting endemic biodiversity (a key component of natural heritage for UNESCO) from endangerment by humans and nonnative invasive species.

Fifty-six years old, Ali was the oldest conservation agent and had worked for the project since 1991. He was stationed in the village of Sahasoa, where slightly offshore was the marine reserve, about 1,000 hectares large, including three islets and a coral-encrusted seabed teeming with marine life, which was partly visible from the surface at low-tide. Each of the uninhabited islets was named for its predominant feature. Germania Tree Island (Nosy Antafaña) was the largest, and its beach was strewed with the tree's large, smooth, sculpted-looking seeds. Flocks of Saunders's terns flew above Bird Island (Nosy Vorona). Rangontsy Island (Nosy Drangontsy), girded by mangroves, was named after three brothers whose bones lay boxed there in the shade of a granite outcrop.

Ali stayed in the village proper with his much younger wife and small daughter only part-time; otherwise he alone or occasionally the whole family occupied a simpler house on Nosy Antafaña for several nights of the week. From there, Ali patrolled the marine park in his motorized outrigger canoe at high tide, making sure that fishermen were authorized members of the village fishing associations. When the fishermen beached their pirogues on Nosy Antafaña at the day's end, Ali and his assistant checked their glistening catches of octopus and fish for undersized or endangered species.

Ali described himself as "severe" with the fishermen and therefore unpopular in the village of Sahasoa, but residents I encountered seemed to like him and find him amusing. He had a gruff, insouciant manner, and was reputed to like drinking beer in the evenings in the small restaurant-bars. His mother was born in a village of the district, on the other side of the mountain, and his father was Comorian, from the island of Anjouan, which explained his Muslim name. His coworkers in the project would laugh when describing Ali to me, such as his strict abidance by the *fady* of the marine park, and the fact that he had a very young local wife—a "child of the mountains" (*zanatanety*), a peasant. Ali was especially vigilant about the rat *fady*. His austere, one-room house on Nosy Antafaña was completely overrun by rats. He had to sling his belongings over the beams of the roof and stored plates and utensils in metal chests weighted down with stones to prevent the rats from ransacking everything. They were incredibly bold, even in broad daylight.

Rats were not usually tolerated in Madagascar; they were considered dirty rodents most of the time. The rats of Nosy Antafaña were special, as were four other species: the skink (a type of large lizard), the tern, the zebu, and the wild boar. Of the five types, only the first three, skinks, terns, and rats, occurred there naturally. Cattle and boars, in contrast, would appear only as meat brought over by visitors, which was *fady*. It was also *fady* to kill, injure, or utter the real names of these five animal types while on the islets, and rather than speak their real names, one had to use euphemisms for them while visiting the islets. The table below lists the Betsimisaraka animal names and the euphemisms one had to use when referring to these animals while on the islets of the marine park.

Failure to use the euphemisms could offend the animal, resulting in bad luck or worse. These prohibitions had been established during the lifetimes of the Rangontsy brothers, the founding ancestors of Sahasoa. What do these euphemisms mean? Were they historical clues, some earlier folk taxonomy? Was the organizing principle forgotten now, or just inaccessible to me, a foreigner? Like the *fady* governing gesture and practice around these animals, the nonsense of the euphemisms distanced the speaker from danger.

Common name		*Euphemism*
androngo (skink)	→	*sahidilahy* (male witness?)
sikôza (tern)	→	*tsy mirirana* (doesn't curve, doesn't lean)
voalavo (rat)	→	*tsy mamaky* (doesn't traverse, doesn't read)
aomby (zebu)	→	*lava vava* (long mouth)
lambo (boar)	→	*antaniloha* (in the original country)

The British Reverend James Sibree (1892) describes the Malagasy custom of designating a euphemism for animal names that were shared by chiefs or royalty:

> The names of the chiefs almost always contain some word which is in common use by the people. In such a case, however, the ordinary word by which such thing or action has hitherto been known must be changed for another, which henceforth takes its place in daily speech. Thus, when the Princess Rabodo became queen in 1863, at the decease of Radama II, she took a new name, Rasoherina. . . . Now *sohérina* is the word for chrysalis, especially for that of the silkworm moth; but having been dignified by being chosen as the royal name, it became sacred (*fády*) and must no longer be employed for common use; and the chrysalis thenceforth was termed *zàna-dàndy*, "offspring of silk." So again, if a chief had or took the name of an animal, say of the dog (*ambóa*), and was known as Rambòa, the animal would be henceforth called by another name, probably a descriptive one, such as *fandròaka*, i.e. "the driver away," or *famòvo*, "the barker" &c. (Sibree 1892:226–227)

It is conceivable that the Rangontsy brothers of Sahasoa had a chiefly status in life, thus taking on several of the animal names in question and compelling people to give them pseudonyms. Moreover, reptiles, certain birds, cattle, and domestic pigs are common *fady* animals in Madagascar. They are vested with a sacred/dangerous status. For example, in the biosphere reserve, Betsimisaraka residents claimed it was *fady* to harm or kill chameleons because whatever one did would manifest in kind on one's person. As for the nonendemic wild boar, this animal was relished by Betsimisaraka but hunted to the point of endangerment in Mananara-Nord, whereas domesticated pig (*kisoa*) was *fady* to eat, doubtless because it was a favorite Merina dish.

The metamorphosis of *lambo* (wild boar) into the phrase "in the original country" intimates a human memory of the flow over space and time of language and species. The memory may be retrievable in language clues. Roger Blench offers a linguistic analysis of the origins of Malagasy animal names, including wild boar, cattle (*aomby*), goat, dugong or sea cow (*lamboharano* in Malagasy, "wild boar of the water"), and others. His findings corroborate zoographic evidence for the translocation of domestic and wild species across the Mozambique Channel and between Madagascar and the Austronesian islands. Nearly all domestic animal names are borrowed "from the languages of the coastal Bantu and Austronesian traces are found only in fossil forms" (Blench 2006:2), suggesting that animals endemic to Austronesia traveled to mainland Africa first, shaping the social history of Bantu populations, and then were brought to the island. Blench continues:

> An intriguing example of this is the Malagasy name for the wild pig, *lambo,* which reflects Austronesian names for "bovine." Given the importance of pigs in Austronesian culture, such a replacement may seem surprising, but it seems that the ancestors of the Malagasy transported very large wild pigs from the African mainland as a food source, and these seemed more comparable to cattle than pigs. In the meantime, the *importation* of mainland cattle brought the Bantu name *ŋombe,* which replaced exiting Austronesian terms. The term *lambo* in turn spread to Shimaore, the Bantu language of Mayotte, where it is applied to the dugong. (Blench 2006:2)

Lambo, a domesticated pig from continental Africa (denoting perhaps "the original country" of the euphemism for boar) had become part of the Indian Ocean trade route, joining the current of African species and persons who landed up in Madagascar. The choice of the word *lambo* for wild boar invokes that history and suggests that the domesticated animal, after its translocation across the Mozambique Channel, devolved to a feral state. Persons translocated from Africa to Madagascar were bought as slaves by Merina elites or sold again overseas.

Connected to the negative actions surrounding the five animals was the positive action of paying tribute to the Rangontsy brothers when visiting their islet. Ali invited me on a daylong tour of the islets. He warned me ahead of time to bring some small change. But I could not find a single coin in Sahasoa village before leaving. Not a single shopkeeper or restaurateur could break a bill for me. Ali told me not to mind. It would be all right.

At Nosy Antafaña, Ali beached the pirogue, and from there, we waded through the shallow tide to Nosy Drangontsy, the site of the tombs of the Rangontsy brothers. Careful to avoid the sharp spines of urchins (*volovava,* "hairy mouth") and the half-buried razor clams (*silasila*), we waded through the umber roots of mangroves. Stepping foot on Nosy Drangontsy, we headed toward the center, where the tombs (*fasaña*) lay. The *fasaña* was comprised of three boxes of bones nestled under an outcropping and marked by a wooden sign. We stood in front of a tomb, and Ali began to make a formal speech, introducing me to the ancestors. He finished and turned away.

Suddenly, then, he spun around to face the tombs and yelled out, "she will learn! She will learn!" I realized by his mood then that Ali had been hiding his anxiety about my failure to offer any coins to the ancestors. At the day's end, we prepared to boat back to the mainland. A few minutes out to sea and the motor died. Ali swore and shook his head, and I could see his fear. The wind was picking up, and waves lapped the sides of the pirogue. We were too distant to row against the tide to shore. I fretted over Ali blaming me for not paying tribute to the Rangontsy brothers, or for accidentally having said aloud the day before on Nosy Antafaña the word *lambo* for boar instead of its euphemism, *antaniloha.* Or he might have blamed me for having blurted "Oh! *Tsy mamaky*!" when a large rat lunged onto the table at Ali's house. Even though I had used the correct euphemism, it was apparently bad to direct any unnecessary attention to the rats, as I gleaned from Ali's stone-faced reaction. Fortunately, the boat's motor restarted after ten minutes or so, and we arrived safely on shore. Later in the evening, as we drank beer together in a small restaurant in Sahasoa, Ali admitted to me he had feared that the ancestors killed the boat's motor because they had been offended. He remembered a time when that happened.

It was in the late 1990s, when administrators of the biosphere project had staged a campaign on Nosy Antafaña to eradicate all the nonnative species and to restore the islet to a (putatively) pristine condition. Biosphere workers at the time had to uproot all cultivars, such as lemon and breadfruit trees, banana plants, and coconut palms. Then they hired a specialist to poison the rats. The animals were baited with wax blocks saturated with anticoagulant poison (Cooke et al. 2004:207). "The biosphere told a big lie back then," said Ali; "many many rats died." Ali claimed that the biosphere bosses had promised him and Sahasoa residents they would leave the rats alone. People feared the anger of the Rangontsy brothers, who might choose to make the waters choppy and unfishable, as well as kill boats' motors leaving one at the mercy of the tides. As proof, the motor of the ICDP's canoe irreparably died, and the project had to buy a new one.

By 2000, Ali and the villagers had nonetheless survived the wrath of the rats, and the species had rebounded spectacularly on Nosy Antafaña. The ICDP had desisted from its rat eradication plans in order to smooth things over with Sahasoa residents, since the biosphere staff did not need another reason to stoke resistance against conservation. Ali, like other Sahasoa fishermen and villagers, not only remained vigilantly observant of the animal *fady,* he also came to track the transgressions of the biosphere administrators.

Abidance to the *fady* of Sahasoa resisted the hierarchy of species value imposed by conservation authorities, asserting a moral code that privileged "cultural heritage" over endemism, or an essentialized "natural heritage." The proliferation of the nonnative, common rat (*Rattus rattus*) on Nosy Antafaña thwarted conservationists' desire to restore the islet's landscape after an image of an earlier time, and presented a continuous constraint to this effort. Ali, who like the other conservation agents of the biosphere reserve project negotiated the perspectives of conservation authorities, who paid his salary, and resident fishermen, who offered him free meals and companionship, resolved that tension of conflicting loyalty by being strict about enforcing conservation rules, on the one hand, and insistent about respecting the *fady* that existed in Sahasoa before the creation of the biosphere reserve.

CONCLUSION

Unlike *in situ* natural history museums, world heritage sites and biosphere reserves do not encase bits of heritage but inscribe them onto lists and in documents. They require labor to valorize the intrinsic value of elements of nature/culture. The preservation of cultural and natural heritage constitutes what Sian Sullivan describes as "a new wave of semiotic and material enclosure of 'the global environment' into a range of derived tradable commodity forms, to produce an increasingly 'derivative nature' of complex, virtual nature products" (2010:14). In so doing, the production of heritage sites and heritable objects, those "irreplaceable sources of life and inspiration," defines places for outsiders primarily according to predetermined criteria of value. Heritage narrows and simplifies diversity for the consumer of this literature; it streamlines history and occludes historical episodes of social resistance to state authority.

Conservationists are compelled to protect biodiversity for the outward forms of biodiversity and habitat, as well as to discover survival resources, such as medicinal properties in the biological matter of rain forest species. Survival potential exists in the relicts of an ostensibly more authentic past, in living-dead matter. Living-dead matter refers to ecologies shaped by human labor. They therefore reflect the products of this labor while also possessing an independent life force. Conservation seeks to eradicate traces of labor and to unleash nature's regenerative capacity.

Through ecological restoration, the re-education of human nature, and the valorization of existing cultural formations that evoke an earlier, more harmonious relationship between humans and the environment than in the present day, conservation labor aims to resuscitate an imagined, ahistorical past. Wilderness, material and external, is imagined to be antecedent to culture (Lien 2005; Ginn 2008). Nature and culture are seen as categorically distinct, and valuable culture within global conservation discourse becomes that which safeguards, or does not diminish, original nature.

If heritage in global conservation discourse appears thinglike in its conservability and accretion of value, Malagasy *fombandrazaña,* in contrast, seems pliable and dynamic. It encompasses a diffuse set of practices and beliefs that keep the past alive while attending to the exigencies of the present. Nature and culture, distinct ontological realms in Western imaginaries presented as separate forms of heritage, are fused in a Malagasy symbolism of descent. By turning deceased kin into active participants in subsistence labor processes, as well as in the politics of conservation and development, the reproduction of *fombandrazaña* makes the dead matter greatly.

7 Cooked Rice Wages

Internal Contradiction and Subjective Experience

Most conservation agents of the Mananara-Nord Biosphere Reserve thought at one time or another about quitting the ICDP. Jafa said in May 2001:

> I'm lazy, tired. . . . The reason for being tired is that the wages aren't fair. It's been three years and the wage hasn't moved, hasn't risen, and the work is hard. It was already only like cooked rice wages. And we're stuck there. (Field notes 5/11/2001)

"Cooked rice wages" (*karama vary masaka*) was an idiom for the bare minimum needed to buy rice for one's household—rice being the foundation of all meals in Madagascar. Since Jafa, like the other conservation agents, never got the raise he expected with the biosphere project, he felt structurally stuck (*tsy mietsika,* "to not move"). The job dashed his ambition and made him feel lazy (*kamo*) and exhausted (*visaka*).

The phrase "cooked rice wages" conjoins two "media of value," rice and money, which are "the concrete, material means by which . . . value is realized" in the subsistence economy and in capitalist workplaces of the eastern Malagasy forests (Graeber 2001:75). Wages (really the monthly salary) equal money, the medium of value par excellence. As David Graeber (2001:66) writes, money is the "very embodiment of value, the ultimate object of desire" in capitalist society.

Money buys rice, among other things, but villagers buy it only when expedient. It is far preferable to cultivate it oneself. The cultivation of rice is made possible by the expansion of landholdings onto fertile soils. Land clearance fulfilled immediate household needs as well as the future needs of offspring, who inherit land from their parents. The more land one possesses in rural Madagascar, and the more family members available to help farm it, the more rice one can harvest for consumption, storage, or sale. Rice affords security and symbolizes the minimal level of well-being for all Malagasy. Even if they had access to other starches, such as manioc or potatoes, eating other starches was not satisfying. For people in Mananara-Nord, not eating rice at a meal was equivalent to not eating a meal at all. The possession of cooked rice at the very least staved off despair. People whose harvests were insufficient to get them through the lean season as the next crop of *tavy* rice ripened saw themselves as destitute, ashamed to be dependent on kin to share meals.

With "cooked rice wages," Jafa thus expressed recognition of his structural position in the ICDP, hovering at the cultural equivalent of the "poverty line" and reflecting the low value of manual labor relative to administrative labor. In suggesting "at the very least I can get rice with this salary but nothing else, so it is not worth the effort," Jafa was justifying his inclination to quit the ICDP so that he could devote all his time to subsistence agriculture and to working in his small store in Varary.

His words exposed a core contradiction of conservation and development effort: that is, lower-tier employees perceive the subsistence economy to offer a better living than wage work. I knew that the ICDP salary was a very desirable income and economic buffer for conservation agents, but what embittered them about the job was that during periods of intensive labor for the ICDP, their social relations in the village were strained, and their households suffered their absence when they could not participate in agricultural tasks. The contradiction of neoliberal conservation and development, and of earlier as well as coterminous forms of forest valorization authorized by the state, was that advocates of the neoliberal paradigm condemned *tavy* yet relied on cheap, local labor. Rural labor comes cheaply thanks to *tavy*.

I have suggested that conservation agents, though comprising a very small labor force relative to the land surface they were hired to protect and produce, epitomized the internal relations of biodiversity hot spots and heritage sites, such as the Mananara-Nord Biosphere Reserve. How did the class antagonisms of the conservation and development bureaucracy contribute to an accumulating crisis of the "second contradiction" (O'Connor 1998)? By this, James O'Connor (1998) refers to the "ecological crisis arising from capital's degradation of its own conditions of production on an ever increasing scale" (Foster 2002). In this final chapter, I describe how conservation agents embodied the deeper contradictions of capitalism and conservation in the rain forest of Madagascar, as well as how they negotiated their lives as walking contradictions of the conservation economy. These contemporary workers played a direct role in protecting the longevity of rain forest species but they also exacerbated biodiversity loss. Instrumental in communicating with and patrolling the erosive actions of peasants, conservation agents were nevertheless "stuck" on a low rung of the bureaucratic ladder, receiving far less pay than the alienated experts who compile lists, attend meetings, and write scientific and summary reports of the nature reserve and ICDP's accomplishments.

The chapter begins with an account of each conservation agent's station in the biosphere reserve and placement within the dual economy of subsistence labor and "conservation and development." I then focus on two periods in which conservation agents felt the pressure of dual loyalties and obligations and expressed anxiety, fear, or self-defensiveness with respect to their ICDP jobs. These periods included the *déguerpissement* (forest sweep) of 2001, when conservation agents joined forces with the national police (*gendarmes*) to patrol the national park and apprehend *mpiteviala* (clearers of forest), and the transition of the ICDP's oversight from UNESCO to the park service, ANGAP, whose representatives, poised to become the new bosses of the conservation project, were deciding which employees to retain or let go.

CONSERVATION'S WORKER-PEASANTS

In Mananara-Nord between 2000 and 2002, I found that the conservation agents spent more time on subsistence labor, vending, cash cropping, and ritual labor, such as preparing for ritual sacrifices of chickens or zebus to the ancestors,[1] than they did on conservation tasks. This was more due to a poor coordination of tasks by the ICDP's national director than it was to defiance on the part of conservation agents.

At the same, the ICDP bosses expected them to serve as model agriculturalists and emissaries of the conservation message when they were not busy doing other concrete tasks for the ICDP. In late 2000, many of the conservation agents complained of a "lack of coordination" in the ICDP. There weren't many tourists visiting the park, so although by 2001 the conservation agents had been named the official tourists' guides of the biosphere reserve, they did not get practice. While the ICDP bosses did not advocate that manual workers carry on their normal village lives when no structured conservation and development tasks had been assigned for the month (which had become an increasingly common occurrence), the conservation agents were not inclined to police their neighbors and kin or promote conservation on their own initiative.

The geographical locations and economic networks of the biosphere reserve's conservation agents brought to light the internal relationship of *tavy* production and conservation in Mananara-Nord. By living as other villagers in Mananara-Nord did, conservation agents essentially "stole time" from the ICDP—to use the language of union workers in the United States—as they made ends meet. Due to the nature of the workplace (remote, mountainous village settlements and rain forest), they could easily cloak their subsistence practices from the eyes of their bosses.

Both conservation agents and development agents were stationed around the reserve in different sectors, each of which possessed one biosphere reserve passage house and office in the main village. Each also had a sector *chef* in charge of a local team of conservation and development agents. The sectors included five villages (Antanambaobe, Sandrakatsy, Anove, Antanambe, Sahasoa), the town of Mananara-Nord (called Mananara-ville), and Nosy Antafaña, one of four small islets in the marine reserve, where one conservation agent was stationed to monitor fishing in the coral reef system.

The ICDP staff anticipated ANGAP's takeover to happen in early 2002. A year earlier, in early 2001, the project had a total of ten conservation agents, as two men had recently resigned, two *chefs secteur* ("sector leaders"), and the leader of the whole conservation crew. Midway through the year, one agent and the head of the conservation crew resigned, leaving eleven in all. The crew was spread thin. Each of the conservation agents and sector leaders was responsible for monitoring his sectoral region. For example, Sylvestre, stationed in the Sandrakatsy sector but residing in the neighboring village of Varary, was responsible for covering Varary, Sandrakatsy, and the land up to the boundaries of

Antanambaobe village. Another agent, Jean-Luc, lived in the village of Antanambaobe. He was thus responsible for the lands around this village. Because of his two years of university, however, Jean-Luc's reach of responsibility also extended to the marine reserve and other key spots where his knowledge in "ecological monitoring" was needed.

Four of ten regular agents resided in villages on the western side of the reserve, "off the beaten path," meaning their villages lay at least seven kilometers from the main road and were accessible by foot only. Three agents and two sector heads were assigned to the eastern side of the reserve. Due to the poor condition of Route Nationale 5, the drive into town often took ten or so hours from Anove village. Like other Mananara inhabitants, sometimes agents made the daylong walk into town if they could not hitch a ride (the director of the project frequently used the working project vehicle for his own needs, making it unavailable to the staff). Likewise, when their monthly wages arrived from the capital, biosphere employees had to either pay the bush taxi fare into town to collect the wages, or else walk. They claimed they were not reimbursed for their taxi fare.

Two of the agents on the eastern side, as far as I could see, spent a great deal of time in Mananara-ville, far away from their stations. Raleva, the agent in charge of GCF (Contractual Forest Management), lived steps away from the biosphere office and regularly took long treks to remote villages to hold GCF meetings.

In sum, the primary residences of three conservation agents were villages, where they regularly farmed rice and other crops. Two agents were stationed in Mananara-ville and made outings to more distant villages for their work. Ali, the agent in charge of the marine reserve, hopped back and forth between the village of Sahasoa and the large islet, enjoying the companionship and generosity of fishermen at the end of the day, when a group of them would share a meal of grilled fish together at Ali's house on Nosy Antafaña. Five members of the conservation crew were stationed in roadside villages and engaged in subsistence farming on land far from the national park, land that either they or their wives owned. Of these five, three spent a large part of their time in Mananara-ville, the residence of their wives and children, and traveled back and forth from their village to Mananara-ville as a function of their household economy.

Stanislas and Thomas, for example, were stationed in Antanambaobe, a western roadside village. They lived in the simple wooden house that served as a passage inn for biosphere workers and an informational center for town residents. Formerly an "agricultural agent" for the development component of the biosphere project, Stanislas was trained in the cultivation of cash crops, including coffee, cloves, and vanilla. He was also experienced in rice cultivars and thus had been in charge of training *paysans* (peasants) in improved techniques for riziculture. He was fifty years old, married, with two grown children, and had come to Mananara-Nord from Fenerive-Est in 1978. His wife lived in Mananara-ville and ran their small store, selling rice, coconuts, and "other things" (such as dried beans, cookies, seasonal crops). The village of Antanambaobe stood a short drive, about forty-five minutes when the road was dry and firm, from Mananara-ville. Stanislas regularly sold rice bought from farmers of Antanambaobe in the marketplace there. Thomas, his colleague, was also married with two teenaged children. He too supplemented his biosphere wages with market vending in Mananara-ville, where he sold greens and small produce.

Serge, a conservation agent assigned to the eastern side of the reserve, in the town of Antanambe, had a wife and two small children in Mananara-ville. His brightly painted house stood on the outskirts of the town center, on a sandy road lined with coconut palms bent like bows from the constant wind. The road led to the tip of the serene peninsula called Ambitsika, a grassy clearance surrounded by a view of the sea and the forested shores of Antongil Bay, where families brought picnic lunches on holidays. Serge, forty, was born in the south, in Vatomandry, and moved to Mananara-Nord just before he began working for the biosphere in 1988. The year before, he had conducted surveys in the region for the World Wide Fund for Nature, the organization that conducted research in Mananara's forests before UNESCO took over. During his early years with the biosphere project, he participated in building dams to facilitate rice plot irrigation, and then became a "forestation agent," assisting with the village nursery in Mananara-ville. After a time, he became a "fishing agent," training fishermen in sustainable fishing practices, and was therefore assigned to the sector encompassing the coastal villages. During my stay in Mananara-Nord, I saw that Serge spent much of his time, before the onset of ANGAP's control of the project, at his home in Ambitsika, where he and his wife tended a small store (*épicerie*) attached to their house. I also noticed one day a fishing seine strung up to dry in his front courtyard, suggesting that he may have fished the waters around the Ambitsika peninsula for household consumption or to sell fresh fish at the market.

Zalahely was another conservation agent who spent a large part of every month away from his official work station. Thirty-eight years of age, and married with three children, Zalahely joined the biosphere project in 1990. He was assigned to work in the village of Ambohimarina, on the western side near the national park. His wife resided there, but he admitted to spending most of his time in Mananara-ville. When I asked if he had land, to know if he supplemented his income with agriculture, he shook his head. "The project wages are my only source of revenue." When I asked if his wife owned land, however, he admitted she did. She had a parcel in Ambohimarina on which they both cultivated rice and other crops. "The wages are not sufficient," he said with reference to the biosphere project, "but on the good side I like [the work] because I earn money."

In Mananara-Nord, most of the biosphere conservation agents had to balance the cyclical demands of farm life with the "phasic" responsibilities of wage work in the ICDP. For their agricultural work, they were, in general, rigorous. For their conservation duties, they tended to exert themselves at critical junctures of the ICDP. These junctures were the transitional periods of contracts ending, or new grant cycles, when bosses and prospective bosses exacted more intense labor from their workers. The time-spaces of conservation—the agricultural seasons of protected areas and the programmatic phases of environmental planning—affected employees at all hierarchical levels. But conservation, as well as development, agents had to coordinate the tasks of both waged and agriculture work. In the Mananara-Nord Biosphere Reserve, the labor structure and process compelled conservation agents to negotiate conflicts of interest in ways that often resulted in some breach of loyalty to their ICDP job or to their kin, or some assailment on biodiversity.

DISCONTENT

Most of the conservation agents had joined the project in the late 1980s or early 1990s, when the establishment of the reserve required intensive labor, and diligence was at a high pitch. After the boundaries of the park had been mapped, tree nurseries built, habitant species inventoried, and trails cut, the ICDP's momentum slackened. A decade later, these same conservation and development agents possessed larger households, and because they were older, they dreaded being called away from the village for extended periods of time, such as during the *déguerpissement* (the forest sweep) or when conservation agents had to check trucks at the gate in Anove village on Route Nationale 5. They complained that their monthly tasks were unclear and uncoordinated. The *déguerpissements* and timber control were risky and took them away from necessary agricultural and household tasks, overburdening their wives and children.

They complained that the project was very slow to reimburse agents for any out-of-pocket medical expenses. Of the two project vehicles, one was in constant disrepair and the other was often absent, being used by the Malagasy project director on his long overland trips south into Toamasina. Project employees stationed in villages far from Mananara-ville had to find their own transportation into town if they wanted to collect their monthly salary. Rather than pay for a *taxi-brousse,* an expense that the project was supposed to have reimbursed but no longer did, manual workers opted instead to go on foot, a trek that took between five and eight hours for agents living in villages on the western side of the biosphere reserve. Agents thirty-five years of age or more complained that they were too old (*antitra*) for this type of work, these long walks through the forest and between villages. Their age, however, did not slow them from working industriously in their rice fields and groves. Especially aggravating to the lower-tier ICDP workers was the lack of raises for the past four years.

As the period of UNESCO's transfer of the project to ANGAP drew near, workers felt compelled to prove their diligence. For one, the project bosses demanded of workers a burst of activity to hand over to ANGAP what looked like a well-managed and accomplished project. Furthermore, workers desired to keep their positions. The problem of contractual work for low-level employees, of being laid off with the end of a project phase, represents one way in which employees' sense of loyalty to their jobs was jeopardized by the lack of security—a security once afforded by civil service jobs in the forest service.

The changeover to ANGAP had implications for the social status of the family as well. In contrast to their contractual employment with

UNESCO, when workers' families did not receive medical benefits, employment with ANGAP promised to be better, including family medical coverage, raises, and regular training. However, ANGAP also required official paperwork. As many of the locally hired biosphere employees had arranged customary marriages, and had seen their babies delivered by midwives in the village, or had not attended school for many years (several stopped schooling at the junior high level, the minimum education level demanded by the biosphere project), their lack of documents provoked great anxiety. Conservation agents scrambled to get formal documentation of their school certificates, marriage licenses, and birth certificates for their children in order to legalize the status of their relations. Obtaining licenses and certificates for marriages and births performed in the customary manner required persistence, patience, and, frequently, monetary gifts to local officials. Several conservation agents found it impossible to get school certificates so many years later. "In 1962, they didn't give out BEPC certificates," Sylvestre complained.

THE *DÉGUERPISSEMENT*

When Jafa complained of "cooked rice wages," he implied that his ICDP pay was incommensurate with the toll taken on his body when he walked for kilometers through the forest on strenuous footpaths. The wages also did not compensate the damage to his reputation in the village after the forest sweeps. Whenever the conservation agents together with the national police (*gendarmes*) arrested rule-breakers, the judges and town mayors would end up letting them go without consequence, he explained. This claim may not have been not entirely accurate, since I had also heard that numerous villagers had been monetarily penalized or made to perform community service, such as cleaning up trash in town, after the last forest sweep of 1999. Whatever the true story, villagers had been angered by the *déguerpissement;* perhaps magistrates had become intimidated by growing discontent amongst the rural hamlets.

Conservation agents dreaded the *déguerpissement* more than any other job duty, but if they refused to participate they would be fired. The *déguerpissement* created bad blood in the village and made it difficult for conservation agents to maintain their "insider" status, their source of social capital for the ICDP. In September 2001, the ICDP organized a big *déguerpissement* to find and remove "delinquents." The sweeps were supposed to take place every year, but popular opposition to the biosphere project in Mananara-Nord pressured the ICDP bosses into canceling the sweeps for three successive years. In fact, the ICDP only had conducted two large-scale *déguerpissements* since the establishment of the biosphere reserve in 1989. The ones carried out in 1998 and 1999 created political fallout as politicians in Mananara-Nord condemned the biosphere reserve for taking away peasants' land and forcefully removing them from their ancestral territory. The ICDP opted to not conduct a sweep in 2000. But in the fall of 2001, as the ICDP staff faced the anxiety of UNESCO handing over the reins of the project to the national park service, ANGAP, the ICDP bosses organized a flurry of activity in the national parks. Teams of conservation agents and *gendarmes*, who had the authority to carry firearms and arrest people, convened at the village of Antanambe, the location of a satellite office of the ICDP. Here they would receive their supplies of food, tents, and rain ponchos for spending the next several weeks in the heart of the rain forest.

The office was a solitary building situated along the arc of the Antongil Bay, with a southerly view of a narrow, verdant peninsula that looked uninhabited and untouched. During the *déguerpissement*, the atmosphere of the biosphere project in Antanambe completely transformed. It felt like a barracks as conservation agents and *gendarmes* in military khaki dined by candlelight on the picnic table outside. Heaping ladles of rice and soup into their bowls, the men talked loudly, charged up, their voices drowning the sound of waves lapping on the beach. They rose before daybreak for breakfast, then split into teams of six men, conservation agents and armed police, to stage descents from different entry points into the national park in search of rule-breakers. I did not accompany the crew for the *déguerpissement* of September 2001. I was in no way physically capable, and it was dangerous. I gave a camera to Raleva so he could record photographs of what he experienced. He looked forward to having a visual record for himself as well. (Unfortunately, most of the photos were ruined during the development process in Antananarivo.)

The teams of conservation agents and military police were charged with rustling up "forest clearers" (*mpiteviala*) and bringing them to court in Mananara-ville, where a judge imposed terms of community service or jail time. This was dangerous work. Peasants were headstrong, the conservation agents complained. They often refused to abandon their newly cleared plots in the forest. They sometimes inflicted curses (*manaña aody*) on the approaching conservation crews by leaving hexed charms on footpaths that the conservation agents and *gendarmes* unwittingly walked over. (Leaving evil charms on footpaths was a typical method of causing injury.) The conservation agents fretted about this afterward, ready to attribute ill luck to a curse. Sometimes peasants would brandish their machetes as the ICDP teams approached their homesteads. Raleva had taken to hiding a cheaply made pistol in his fanny pack in case of assault. The forest was dark, the men told me later, sometimes impenetrably dark—a "total eclipse!" in the words of one crew member, alluding to the media blitz around the total solar eclipse a few months earlier.

The long recess since the last *déguerpissement* in 1999 had emboldened rice farmers to clear land in the reserve's core. Conservation agents took issue with the bosses' decision to conduct the sweep in September, when peasants had either already cleared land or were in the midst of burning. Damage to the forest had already occurred, and arresting farmers after the fact only provoked the population. Delinquents who had been apprehended during the previous *déguerpissement* were sent to court in Mananara-ville.

Conservation agents complained about the double-talk of Mananara politicians with respect to enforcing penalties. When the last *déguerpissement* took place in 1999, for example, local officials did not enforce conservation laws or sentences against rule-breakers. In 2001, Etienne had worked for the biosphere for ten years. He found the most enjoyable aspect of his ICDP job to be "protecting the forest" (*fiarovana atiala*) because "there are little children who will live to see it," he said, reiterating the main principle of conservation as it was taught in Madagascar. He did not like policing the forest and resented the fact that the agents' labors seemed wasted by the lack of follow-up by town authorities. "I don't like the corruption of the state," he said. "Reports are falsified, especially at court. People are let off. Nothing changes. The forest clearers don't get enough punishment. That's what makes me bitter."

In theory, rule-breakers either were fined or, if they could not pay, were given a term of community service in Mananara-ville. The penalties were differentially and arbitrarily enforced by magistrates. What was worse for those apprehended was losing the hard-toiled plots of land they had cleared. Even if peasants clandestinely returned to their clearings later, which was often the case, they had already lost workdays for essential tasks in the farming cycle.

Villagers considered conservation agents' participation in the *déguerpissement* a betrayal to kin and kind, and the social repercussions of the forest sweep for agents back in their villages made life very difficult. Serge, an agent based in Mananara-ville, explained how villagers treated him in the aftermath of the first *déguerpissement* as a complete outsider:

> There are those who are lazy [*kamo*] to leave the forest interior. Half of them [*ny sasany*] are not that resentful but half of them are really angry at the Biosphere workers. If a lot of people manage to cut down the forest, then a lot of these forest clearers get penalties [*sazy*]. . . . The half who are really angry—they'd refuse me a drink of water. (Field notes 10/02/2001)

Jafa admitted that people from Varary shunned the small shop he ran with his wife during his participation in the sweep.

I had heard from another ICDP staffer that most Varary residents had been reported for illegal forest clearance during the first *déguerpissement*. This meant that most of the village had participated in deforesting portions of the national park, a fact that Sylvestre and Jafa pretended to not know. Anger toward the conservation agents from their neighbors in the village troubled conservation agents yet it also made them feel righteous indignation—at least as long as they were working with the gendarmes.

I was impressed by Jafa's and Sylvestre's dismissive attitude regarding their neighbors' anger toward the biosphere staff after the *déguerpissement*."They know the rules," Sylvestre said smugly. He appeared genuine, as though he had never bent the rules or broken them in the recent past. I specifically remember a day in March 2001, when I accompanied Sylvestre into the national park for the first time. As we ascended a trail from Varary, our view of the village and surrounding lands widened until we gazed upon a panorama of shorn mountains and valleys with scattered clumps of forest. I thought those were the forests of tombs. But Sylvestre claimed that the stands were remains of the park's earlier limits and not burial grounds. Sylvestre confessed to me inadvertently back then that Varary residents did indeed clear the protected forest.

Our path led us into the cool woods, where we passed green-and-yellow signs constructed and staked by the conservation crew. The signs, in French, noted the distance to a waterfall, the Latin name of a massive thicket of bamboo, and the elevation of the highest peak. Clove trees and red bands of paint around the tree trunks of one side of the trail marked the boundary between buffer zone, in which we walked, and the official park.

To my eyes, the forest closest to the park's buffer zone looked parched, monotonous, and thinly populated with trees, like a newly planted suburban woods. No rivulets or creeks, no flowers. In one area, a stand of trees stood tall and leafless. "What killed these trees? Fire?" I asked. Sylvestre nodded. Here and there, the steep narrow trail was strewn with reddish wood chips. There was some life: birds, enormous fire-orange millipedes (*ankodiavitra*), and giant, round, green beetles (*taimbintaña*). Somewhere far away, the melodic croons of the indri lemur (*babakoto*) resounded through the forest canopy.

We continued, and suddenly heard the sound of chopping that led us to a young couple. The woman sat at the edge of a thick, fallen *nanto* tree (*Sapotaceae*) nursing her baby, and the man, clutching an axe, stood on a blanket of wood chips. He was hacking away at the outer bark of the fallen tree to get at the durable center, the heartwood (*teza*), which was good for building. He greeted Sylvestre, and the two exchanged friendly small talk. I asked the man if I could take their photograph and promised to give him a copy, to which he happily consented as long as I did not reveal him to the biosphere bosses. We then left the couple and ascended a crest on the trail at the forest edge. Sylvestre scared up a flock of guinea fowl (*akanga*) into the treeless fields and explained to me that the man was not felling wood in the reserve proper but outside its boundaries. Contrary to what I knew to be the case, he said that Mananara's national park had no buffer zone, so there had been no reason to reprimand the woodcutter. Everything beyond and interior to the red-banded trees along the trail was protected, he gestured, and everything exterior to the border was open to exploitation. Sylvestre did not seem too concerned whether I was convinced of what he said. I knew that biosphere project maps did delineate a buffer zone in which extracting timber was prohibited. More important to Sylvestre, however, was maintaining good relations in his village. To report on this young family collecting wood in the buffer zone for fuel and construction would unnecessarily create tension over a negligible infraction, some might say. Penalizing the man for cutting a tree in the national park, although blatantly illegal, was more trouble that it was worth to Sylvestre, even in front of me. Maybe his firmness with his neighbors after the *déguerpissement* was a form of self-defense, to shore up his resolve to do his duty until they were done.

The blatant resentment of peasants and defensive righteousness of conservation agents and *gendarmes* that were stirred by the *déguerpissement* faded slowly, but grudges accumulated, particularly of those who were penalized by the biosphere project and state authorities. They were forced to go into town to resolve things. Although Jafa and Sylvestre resumed their old life in Varary afterward, Jafa's store suffered a drop in business, as I mentioned. Sylvestre, who was better off because he had a larger and closer family who mutually supported each other, seemed more inured to the repercussions of his involvement with the *déguerpissement,* perhaps because he was already in a position of authority within the village due to his father's position as *tangalamena.* But he did worry about others' envy and their potential for practicing sorcery against him.

THE TRANSFER OF THE ICDP

The other matter that troubled conservation agents was the impending transfer of authority over the project to ANGAP, an unknown entity insofar as the staff had no idea what to expect from the new bosses. From November through December 2001, biosphere conservation agents were granted a furlough as the project administrators prepared to transfer authority from UNESCO to ANGAP.

Although none were too pleased with the current Malagasy director of the project, he had been lobbying his staff to support his retention as national director. Sylvestre confided that the national director had promised to retain all current conservation agents if he himself was retained by ANGAP. In this way, the director discouraged workers from lodging complaints about him to ANGAP administrators. The director need not have feared disloyalty from the workers, however. The conservation agents expressed no desire for a new director, preferring to stick with a known quantity in spite of his shortcomings. As the biosphere's field agents fretted over whether they would be retained by ANGAP, many residents were anticipating, wrongly, the end of the biosphere project. ANGAP planned to change the men's job title to *agent de conservation et education,* again asserting the importance of the environmental education and *sensibilisation* (conscientization) of villagers in order to establish sustainable land use practices.

In spite of their promotion of the people-friendly approach, ANGAP representatives conducted interviews with and quizzed the ICDP staff in a way that sent a different message in one respect. Conservation agents were questioned about their loyalty to kin. Sylvestre said he had felt let down that the interviewers failed to ask him about his ecological knowledge or his thirteen years of experience as a conservation agent. The interview question that impressed him the most, although he did not believe they were fully sincere in the asking, had been the following: "If your own father were clearing forest, would you report on him?" "Yes!" Sylvestre had answered with momentary conviction.

Ultimately, the tests ended up being easier than expected. This was both a relief and an anticlimax to the conservation agents. "They asked stupid questions," Sylvestre declared. "What does ANGAP stand for?" had been one question. "There are natural and artificial limits to the forest. Explain what an artificial limit is, in French," was another. Agents who knew passable French were told to respond in French, while others with lower levels were not required. They asked Sylvestre, "How many square kilometers was the surface of the terrestrial park?" He found this question stressful because he knew the answer only in hectares, so he had to calculate in his head, "ten hectometers to a kilometer, 23,000 hectares . . . It was difficult, but they accepted my answer." Jafa recalled his own test questions, a few of which had to be answered in French, such as "What does PDC stand for?" Jafa had correctly answered, "Projet de Développement Communautaire." They asked him what, during the past ten years, his tasks had been as a conservation agent. Another question, similar to one Sylvestre had been asked, was "If your work requires that you have your brother jailed, if he was clearing forest, would you do it?" "Yes! One must!" was Jafa's answer. Sylvestre confessed his disappointment and frustration at not having been asked important questions such as his opinion on how to improve the job and its work.

With the reconstitution of the conservation crew, eighteen men were now dispersed among six sectors, and two female conservation agents—in reality environmental educators, the male conservation agents insisted—remained stationed at the libraries of Mananara-ville and Antanambe to offer classes to children in environmental education. The conservation agents scoffed at the change in job title for the women, believing that being a conservation agent was a man's work because it involved potentially dangerous confrontations with peasants in the rain forest, long treks, and heavy lifting. They underscored that the women would continue their teaching tasks rather than joining the manual crew.

The men were pleased with the increase in benefits with ANGAP's takeover of the project. Not only did they receive modest raises, but they also obtained medical benefits for their entire families. Proof of kinship in the form of official marriage licenses and birth certificates was required, however, which caused conservation agents to scramble for official paperwork during the last weeks of UNESCO's tenure. They also received per diems of 12,500 FMG (approximately US$2) for work that took them outside their designated sectors, and a maximum length of twenty days for forest patrols (in the past, some agents were obliged to exceed three weeks' time on patrol).

In addition, the project bosses sought to resolve the shortage of tourist guides by delegating the role to conservation agents who spoke some French. The men faced this duty with some trepidation. All but two men of the conservation crew felt their language skills to be sufficient to guide tourists, and they complained of never having received the guide training they were promised by the biosphere project.

Just before my departure in February 2002, the agents were waiting to implement the start-up tasks. The initial phase of ANGAP's takeover was reminiscent of the hustle and bustle of the first years of UNESCO's tenure. ANGAP's plans would in fact replicate what UNESCO had done during the establishment of the biosphere reserve: a new floristic and faunal inventory of all sectors in the reserve, a new nursery, a possible reforestation campaign. Stanislas explained: "We don't yet know if there will be a yearly *déguerpissement.* We must do a summary report after the first round of *sensibilisation*, patrol—then we'll do a report on the delinquents—a survey, then the legal proceedings." Living in the biosphere reserve meant one was located at the margins of the organizational world of conservation and development, where plans were opaque. Serge, the biosphere conservation agent, reflected:

> Maybe it will be a surprise for people that the biosphere is not gone. They may be surprised to see that the activities are continuing with new employees. At the end of different phases, the people and the politicians said "it's the end of the biosphere!" Perhaps people thought it was over. Perhaps they didn't understand that ANGAP would take over. That is what brings a lot of forest clearance: when people think it's over. . . . People in the forest don't really know what's going on. The state does, but the masses don't. They are far out in the forest. (Field notes 9/18/2001)

Conservation agents who could speak a sufficient level of French (a handful of agents among the new crew) would acquire the role of tourist guide, a role that they relished for its income-making potential but also feared since most of them felt unversed in *vazaha* ways. Serge, but not his family, would relocate to Sahasoa to work with Ali in the marine reserve, teaching the fishermen's association modern fishing techniques, patrolling the waters, and monitoring catches. He surmised there might be more regular patrols of forests, along with *sensibilisation* with farmers about conservation rules and techniques. He was not sure how the conservation agents would go about raising the awareness of farmers, however, and thought the project might provide them training in how to proceed. "There's got to be training," he said, unconvinced it would happen.

With ANGAP's takeover, Raleva, who I felt showed real commitment to his GCF work and the goals of conservation, had seen his wages cut in half to match those of the other agents. Technically, Raleva could not serve as a sector leader since he had never obtained his "baccalaureate plus two years," as had Jean-Luc, who had the "bac" in addition to two years' university education. Despite his lack of formal educational credentials, Raleva had sought relevant training on his own, not through the biosphere project, in communication, adult education, participatory development, grant writing, management, and election observation (to monitor voting during the 2001 presidential election). Raleva had focused his nonconservation practices on increasing his value as a conservation worker, but his efforts could not overcome the structural barriers of the ICDP.

I visited his house near the Biosphere project office in Mananara-ville in January 2002. He had been laid up at home for weeks, his legs swollen and aching from what he said was rheumatism, or a "problem of his veins." Each episode of this rheumatism would trigger a malarial fever as well. He first experienced this disability in 1996. He thought it may have been provoked by the long treks through the forest. "I need to get treated in Tana," he explained, "but the paperwork with ANGAP isn't all set up yet for me to get my medical expenses reimbursed." He could not afford to pay out of pocket, so he suffered in bed, at home. To his disappointment, he had also been transferred to the Antanambaobe sector, where the project bosses expected him to relocate even though his children were in school full-time in Mananara-ville. Since his wife worked in Antananarivo, Raleva took care of his children by himself. "They're my future," he said. "I can't spend three months in the forest doing patrols any more. If that's what ANGAP wants, I'll have to quit."

While most of the conservation agents appeared to look forward to the reinvigorated work schedule, it remained to be seen whether the full load of start-up tasks could be maintained indefinitely. Or would the time-discipline of the project cycle repeat itself, reducing the role of conservation agents to periodic forest policemen who cited and arrested peasants? Sylvestre, the agent who lived in Varary village, was shocked and devastated (*vaky lôha,* "broken-headed") that ANGAP would relocate him from his home village to the other side of the reserve, in Antanambe. The director tried to "fix his mind" (*manambaotra saina*), to console him by promising his station would not be permanent. But in the meantime, his disappointment made him more candid than usual about his job. Sylvestre asserted that among all the biosphere employees, conservation agents "work the hardest." When I said agents should be able to do other things, like tree planting and species inventories, instead of just *déguerpissements,* Sylvestre agreed: "It's like nothing is planned. No program, just the *déguerpissements,* which agitate the population and do not promote awareness about the environment." He continued, "In the beginning, we were doing *sensibilisation* every month, but it has had no effect. And the law isn't supported by the state. The work we do is not supported." He worried that ANGAP had not yet mapped out a work plan for the agents, nor had their monthly wages arrived. He dreaded walking seven hours from Antanambe, his new station, to Mananara-ville every month to collect his wages. He complained that it was also too far, too difficult, for tourists to enter the park from Antanambe. Tourism was not lucrative in Mananara-Nord, he argued. "They need to fix the road, make better trails." Bitter about his job and prospects for improvement, he predicted that given people's refusal to comprehend the consequences of forest clearance, "in ten more years, the little bit of park that remains will be gone."

CONCLUSION

The phases of development programs that are implemented by NGOs for renewable but finite terms create a destabilizing state of suspense or "deferral" (Stoler 2008:193). Ann Laura Stoler (2008) argues that these limbo states are the aftereffects of imperialist domination; "imperial formations" denote the residual impacts of empire. Racialized hierarchies and their skewed apportionments of rights, privileges, and resources persist, yet within a more opaque architecture of power: not state but "parastatal" agencies, like ANGAP.

Under neoliberal conservation and development, governed by agencies and organizations that have been either created by donors or funded by them, subaltern laborers' position fostered sentiments about conservation that worked to undermine the mission. In ICDPs, manual workers' low wages relative to task difficulty, a lack of voice in work meetings, the reneged promises of bosses to provide training that would enhance their skills, the poor coordination of tasks, the denial of medical benefits to family members, and a sense of job insecurity made them feel undervalued and reluctant to abandon the moral economy of subsistence.

The difficulty of disciplining the time of manual conservation labor in a protected area resulted in long stretches of downtime in the ICDP's operations. One cannot literally be in the process of "conserving" nature or proselytizing conservation all the time. By default, conservation agents lived much as they had done before the arrival of ICDP. Their dual existence as forest guards and degraders, *vazaha* collaborators and detractors, presented moral dilemmas and forced them to make pragmatic compromises that had the overall effect of punctuating the rate of rain forest erosion but not averting it. Stealing time from conservation labor furthered the erosion of the rain forest, while *déguerpissements* and the community forestry initiative slowed it down.

Low-wage labor was put in a self-incriminating position of upholding conservation rules at certain moments, then transgressing the rules themselves or turning a blind eye to the actions of others. Rather than separating wage workers from their subsistence practices, a step in the process of economic transformation toward capitalism, or "primitive accumulation," according to Marx, the conservation economy has perpetuated conditions in which Betsimisaraka workers remained attached to the land and thus attached to the social relations of *tavy* and other familial labors.

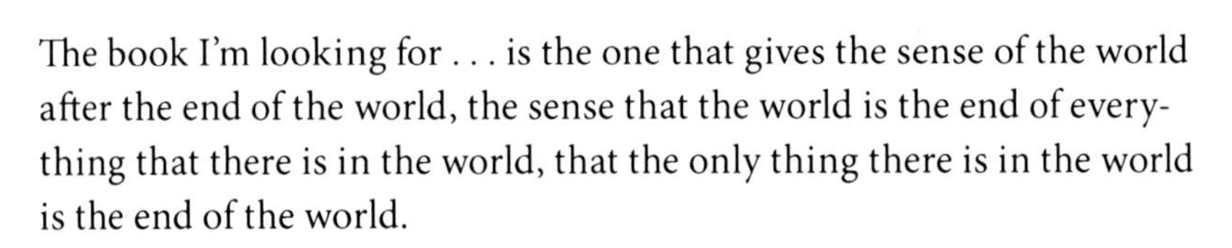

The book I'm looking for . . . is the one that gives the sense of the world after the end of the world, the sense that the world is the end of everything that there is in the world, that the only thing there is in the world is the end of the world.

—ITALO CALVINO, *If on a Winter's Night a Traveler*

Epilogue

Workers of the Vanishing World

As I write these concluding remarks, Madagascar remains in a suspended state of "transition": a presidential election to legitimate state officeholders has not yet occurred. Not seeing an end to the political turmoil, the international political community has resigned itself to recognizing Andry Rajoelina, the usurper of Ravalomanana over two years ago, as interim president. State authorities continue to abet the illegal timber operations, despite certain officials' public statements of disapproval. Conservation activities remain paralyzed there by the threat of violence, but elsewhere in Madagascar they forge ahead, always up against powerful forces of industry from the West and East. The forward momentum of strip-mining, logging, and palm oil production quickens time and shrinks space, as species, habitats, and conservation paradigms fade away, and social life adapts to the new terrain. The extraction of rosewood, teak, and ebony from Madagascar's national parks by Chinese and Malagasy exporters invites comparison between a rapacious, old-school brand of capitalism in the eastern forest and the relatively salutary form undertaken by the United States and European agencies under neoliberalism.

This book has focused on forest-based labor in eastern Madagascar at the twilight of "integrated conservation and development" there. To illuminate the history of Madagascar's transformation from "naturalist's paradise" with countless unnamed species to global "biodiversity hot spot" with countless species destined to extinction, the book explores historical social relations of labor that have endured into the present. My concentration on the low-wage, manual workers of the conservation effort is meant to give some insight into the immensity of the problem of deforestation as well as into the subjective experiences of subaltern labor through time, experiences that have informed understandings of human and nonhuman nature. The labor that has produced the architecture of Mananara-ville and surrounding villages, deforested the mountains, buried the dead and offered them tribute, plucked crystal rock from the earth and octopi from the coral reefs, sown the *tavy* fields, laid the railways and roads, cultivated and harvested cloves and vanilla for shipment around the world, and on and on, has unfolded within hierarchies established by *vazaha,* the mutable outsiders.

DIVISIONS OF LAND AND LABOR IN MADAGASCAR

Under Merina rule, the social relations of production in eastern Madagascar were structured by a racialized caste system—including royalty, free persons, and slaves—that informed the tribalist-racist taxonomy of French colonial officials, who assessed Malagasy populations primarily in terms of the quality of their labor. In spite of the fact that Europeans held pejorative opinions of most of the island's populations, they were enamored of the biological wealth of trees, animals, and plants, as well as the island's geological features.

In 1920, Guillaume Grandidier, a French geographer, ethnologist, and zoologist of Madagascar, proudly declared Madagascar "La France Orientale" (France of the East), reflecting how after a couple of decades Paris viewed Madagascar as a jewel of the empire (Grandidier 1920). The plentiful natural resources of Madagascar compensated for the shortcomings of its people, in the eyes of colonial settlers. Guillaume was the son of Alfred Grandidier, a wealthy industrialist and more famous geographer, zoologist, and ethnologist of Madagascar. Alfred, the father, was a key author of the thesis that prior to human settlement, the island had been blanketed by thick, verdant forest and that Malagasy settlers had in a relatively short time laid the forest to waste. This myth, now discredited, justified the initiation of a forest conservation effort that banned subsistence agriculture in the forest so that the state and colonial settlers could extract raw materials for commodity manufacture. A little more than a hundred years later, la France Orientale had become a Land of the Lost, not solely in the sense of extant, primordial creatures that had resisted the forces of evolution but also in the sense of species that did not survive the livelihood pursuits of human beings.

Another twenty years after independence, in the 1980s, neoliberal reformers ushered in the model of integrated conservation and development, introducing a form of forest-based capitalism based not on resource extraction but on ecological virtualism: the creation of value from intact landscapes and from elements defined as national and global heritage. Toward this effort, an extensive labor force was mobilized to delimit and defend nature reserves, amass and circulate knowledge about Madagascar's rare and fragile biodiversity, educate the will of rural populations to protect this heritage, impose new concepts of space and time on Malagasy society, and strive to meet tourists' expectations of the outward appearance of the rain forest, as well as, to a lesser extent perhaps, the outward expression of Malagasy culture.

Antananarivo, the sprawling highland capital, became the hub of the national environmental program where planners composed the elements of the new discourse: *aire protégée* (protected area), *patrimoine naturel* (natural heritage), *biodiversité,* and *tontolo iainana* (the lived-in world). They circulated these concepts in environmental brochures, posters, comic books, exhibits, and radio and television announcements within Madagascar, on the Internet, and in conservation organizations in Europe and the United States. Expatriate and national consultants, technical experts, grant managers, contractual conservation project managers, ministry officials, agency representatives, national park directors, institutional coordinators, and NGO founders participate in a market of environmental abstractions, removed from the tactile sensations of forest. Nature is available "as a conceptual abstraction, not as a physical partner or opponent in the work process" (Smith 1990:42).

In the discourse of conservation and development, the positive value of biodiversity, entailing an abundance and multiplicity of endemic species that preexist in nature but must be rehabilitated and preserved through conservation actions, gains relevant "potency" by the tug of a negative force (Munn 1986:9), in this case, biological homogenization. This is the result of the degradation of nature: the reduction of plant and animal life in secondary-growth forests, and the colonization of a habitat by alien invasive species, such as the sacred rats of Nosy Antafaña, which outcompete endemic species.

The division between mental and manual labor for biodiversity conservation has corresponded to the island's core/periphery geography. The bureaucracy for environmental protection is centralized in the highland capital, and protected areas are dispersed along the coastal periphery. One could argue that the geographical distance of planners from the physical rain forest is simply an administrative convenience, but I have sought to reveal how the pragmatic aspects of location are tied to cultural conceptualizations of nature. These have evolved out of modalities of travel and perception in the precolonial era between Malagasy elites and subalterns, as well as European travelers and Malagasy porters. These tiered arrangements, echoed at different geographical scales, reverberate in the present day and shape the allocation of certain pay scales and tasks. This spatialized division of labor has furthermore had an ethnologically temporal dimension: the Merina capital of the high plateau, whose distinctive architecture inspired the metaphor "la France Orientale," represented the acme of modernity in Madagascar, while the downward slope toward the coasts sent one backward in time toward ancient forests and more "primitive" social groups. For Betsimisaraka populations on the east coast, the trajectory of slave raiding and colonization came from the "white" center and traveled toward the "black" coasts. "The forest is being protected for the *vazaha*," residents of the biosphere reserve often said to express their view that conservation was neocolonialist.

THE VALUE OF A SALVAGED NATURE

In biodiversity hot spots, the value of protected areas is mediated by foreign exchange, discourses of heritage, extinction, and sustainable development, and compilations such as the IUCN Red List of Threatened Species and UNESCO's list of World Heritage sites. Protected areas are conceived as both global public goods that provide essential ecological services and resources, and quasi-commodities insofar as they are spaces of utility ("natural capital") transformed by the social relations of labor into tourist attractions, objects of philanthropy, and essences of national identity. Like any geographical formation, hot spots and protected areas are social relations.

Marx (1977:165) argues that commodities are fetishes, or "sensuous things which are at the same time supra-sensible or social." The commodity as fetish reflects the social relations of labor as merely "the socio-natural properties" of labor's products (Marx 1977:164). Because they appear to have an autonomous existence, lives of their own, commodities mystify the conditions in which they were forged. Although natural things are vital in the literal sense, natural spaces like parks and untrammeled areas of wilderness are fetishized in Western society. The literal vitality of the protected area, commodified as a tourist attraction, works to establish the existence of nature's intrinsic value. Environmental curricula and media promote the idea that nonhuman species have value in and of themselves, for their own sake, regardless of their worth in a market (Humphrey 2002). Animals, plants, and entire ecosystems are conceived as morally considerable; their intrinsic value thus confers rights upon them, particularly the right to life (Light and Holmes 2002). While one can make the claim that habitats, plants, and animals exist autonomously from human society, one cannot claim that their intrinsic value does. The liveliness of the protected area and environmentalists' recognition of the intrinsic value of wild species shroud the social relations that have forged these particular ideas about nature, such as the view that conservation is essentially a form of negative action rather than value production.

In Madagascar, subaltern labor and the ambulatory knowledge it possesses have facilitated the global perspective of biodiversity and ecological hot spots that underwrites conservation. Just as women's domestic labor and knowledge have historically enabled male productivity in the formal economy and so-called public sphere, thus enabling an accumulation of the products of male labor, manual workers in Madagascar's rain forests have facilitated the European's abstracted, aesthetic relationship to rain forest habitat. There is some irony here, given the view of conservation agents in Mananara-Nord who felt that forest surveillance and park maintenance were clearly men's work. The artful work of enhancing the scenic views in rain forests, or building trails and guiding tourists to beautiful spots served to foreground nature, masked artifice, and as a result diminished in value the conceptual category of labor itself.

Worker-peasants in the rural capitalist economy, including contemporary conservation agents, have moreover enabled foreign conservation and development projects to operate at relatively low cost due the invisible subsidy of rice cultivation. This remained true for ICDPs, since conservation agents were unwilling to abandon the subsistence economy. Conservation agents felt anxieties under the new system of contractual NGO work and so felt compelled to nurture rather than jeopardize their ties to people in the village. With the advent of NGO work, Malagasy workers who had been employed by the forest service saw the security of civil service under socialism fade from view. A more general trend in state rhetoric was a silencing of talk about the peasantry and the proletariat as agents of social change.

The failure of socialist policy to pull Madagascar out of the downward spiral of a global recession opened a space for neoliberal reformers to impose a new value hierarchy onto Malagasy society. Of paramount importance now were free trade, structural adjustment, sustainable development, grassroots participation, and the protection of biodiversity. The category of "labor" had permutated under neoliberalism to "people," "grassroots," and traditional "culture." Labor, unions, proletarians, "the dignity of man" no longer had positive valence; these terms were obsolete and appeared to play no part in the modest annual achievements of ICPDs or in giving social existence to other valuable objects—such as nature reserves, lemurs, sea turtles, orchids, and rosewood. If peasants have added value to Madagascar's heritage, the process of valorization is understood to have been effected through negative means—that is, through peasants' agricultural labor, habitats have eroded and species have become scarce. These contradictions, displacements, and vanishings appear to have nothing to do with the so-called intrinsic value of endangered flora and fauna. Yet, I emphasize, the social relations of rain forest conservation have served to perpetuate the conditions that cause species scarcity.

The protected area is not only a product of intentional labor but also the unintentional by-product of historical labor relations. The interrelationship between the production and by-production of the nature reserve is starkly revealed in the subject of the conservation agent who up until 2009 actively (though sporadically) surveilled the rain forest and protected it by catching and penalizing rule-breakers, yet also eroded the rain forest by practicing *tavy* or allowing, even depending on, others to do so. Conservation agents' willingness to turn a blind eye to rule-breaking was justified by a relatively low salary and low value in the ICDP, whereas their rigor in carrying out clear and coordinated tasks for the ICDP was motivated by the desire to earn a wage. The position of conservation agents, the forest *ouvriers* of the late twentieth century, sheds light on how the intentional and unintentional degradation of forest has resulted in a heightening of the value of particular ecologies within the global conservation economy. Conservationists are not the only ones who impute high value to endangered species; so do those who seek to profit from the illegal traffic in exotic, endangered species. Ironically, the discovery and "red-listing" of species also puts them in danger of capture for trade markets in exotic animals and plants (du Bois 1997). Courchamp et al. (2006) hypothesize that the disproportionately high value placed on rare species could result in "the anthropogenic Allee effect," a term describing a "cycle in which increased exploitation further reduces the population size, which in turn increases its value and ultimately leads to its extinction in the wild" (Hall et al. 2008:75).[1]

To conclude, I will say that despite the critical perspective presented in this book, I share a desire to salvage rare life from oblivion. I embrace global conservationism in principle without denying the dangers of its imperialist impulse, its ideological roots in exploitative labor regimes, or the tendency of many within the ranks of its adherents to impugn the actions of other people in faraway places. Yet I recognize that my own nostalgic perspectives of tropical nature, and my interests in the natural histories and survivalist narratives of vanishing worlds, have also evolved out of this history.

GLOSSARY OF MALAGASY WORDS

adimbilany enough to fill a pot (metaphor for minimum wage)
akanga Guinea fowl
ala fady sacred forest; site of familial tombs
aleo maty amaraiñy toy izay maty iniany "Better dead tomorrow than dead today."
ambanivolo/ambanivôlo countryside; outback
ambava by word of mouth
amboa dog
ambôra type of large tree (*Tambourissa* sp.) formerly used by Betsimisaraka as coffins
anakolokoloko a type of wild plant with small purple flowers named after the sound of a turkey gobbling
angatra ghost
angivy a bitter green fruit resembling a tomatillo
ankôba garden or grove
ankodiavitra millipede
antaniloha euphemism for "boar" on islet of Nosy Antafaña
antitra old, elderly
antsantsa shark
aomby zebu; cattle (Betsimisaraka)
Aza matahotra . . . Mino fotsiny ihany "Do not fear . . . Only believe." Passage from Mark 5:36 in the New Testament used by Marc Ravalomanana during his presidential campaign
babakoto Indri lemur
bakôra a trumpet-shell horn used to convene people to the village
bakose a gait used by porters consisting of elongating the stride and advancing with outstretched legs
baliña dance party (Fr. *balle*)
banansandrata a disagreeable gait used by porters resembling a limping horse that gave the impression that one was going to crash and fall with each step
baromeny a meter-long metal rod used for searching crystal
betsa fermented sugar cane juice
betsimisay a gait of porters that employed an uneven step and resulted in passenger fatigue
biby animal
bisikilety bicycle
bôka leprosy
borizano porter of luggage or cargo in the nineteenth century
eponzy sponge; foam used for bedding (Fr. *éponge*)
fady taboo; proscription established by ancestors or elders
fahahôva period of Merina rule
fahavazaha period of French rule; colonialism
falafa split-bamboo siding used for walls in rural houses
famòvo "the barker" (euphemism for *ambóa* [dog])

fandròaka "the driver away" (euphemism for *ambóa* [dog])
fangatahaña lalaña "request to pass through"; part of the ritual performed to enable the cultivation of land lying next to tombs (*fanómo*)
fanómo land lying next to tombs; its cultivation requires the performance of a special ritual
fanompoana a system by which individuals perform obligatory labor for the sovereign
fasana/fasaña tomb(s)
fiaroaomby "protection from cattle," or a small fence of quickset trees planted around a buried coffin
fiarovana atiala forest protection
fihavanaña familial relations; an ethos of treating village members like kin
filanjana the Malagasy palanquin (raised seat carried by porters)
fisokina a wooden stake on which evidence of sacrificial offerings (zebu horns or chicken heads) to the ancestors is attached
fitoana time
fokonôlona an administrative collective of adjacent villages
fokontany an administrative unit of a hamlet
folera flower
folo andro obligatory "ten days" of labor for public works required of citizens after Madagascar's Independence
fomba manner; custom
fombandrazana/fombandrazaña ancestral customs, traditions
hasimbola type of quickset tree used for marking boundaries around land or burials
hasina/hasiña generative
hazo coffin wood (***hazo*** means "tree" in the Merina dialect)
hazovola species of palisander tree
hintsy species of tree called the "teak of Madagascar" (*Intsia bijuga*)
hôngotra/tongotra foot or leg
hôraka rice paddy
hôrita heart of the ravinala palm
Hôva Merina person (Betsimisaraka); amongst the Merina, *hova* refers to persons of free but not royal status
Iray Aina "One Life," an evangelical workers' association
jinja swidden plot (Betsimisaraka variant of *tavy*)
jiny evil or capricious spirits (Merina)
jirofo cloves (Fr. *girofle*)
kabary formal speech, delivered at rituals or to welcome visitors
kabone lavatory, toilet (Fr. *cabinet*)
kakazo tree (Betsimisaraka)
kalanoro/kalañoro a type of spirit described as short, hirsute, and fanged that establishes a relationship with a living person and imparts clairvoyant powers on him or her or that does the bidding of its human owner
kamo lazy, sluggish, unwilling to work
kapoaka a standard measure used in Madagascar equivalent to a five-ounce tin can
karama vary masaka "cooked rice wages" (metaphor for bare minimum wage)
kavilava moonshine alcohol
kermesse all-night dance party set up in villages by traveling DJs who bring generators
kesikesiky type of quickset tree used for marking boundaries around land or burials
kiranily plastic, gel sandals that are good for walking the rain forest terrain
kisoa pig
lamba cloth, clothing
lambo wild boar
lamboharano dugong
lasa adaladala to go crazy
lava-bato natural cave or dug-out earth, sealed with stones, where the dead are interred
lô ambôra an ***ambôra*** tree that has rot-

ted, exposing the bones of a common ancestor. This duration is roughly equivalent to the passage of nine generations, the moment at which collaterals of the male ancestral line may marry without fear of incest.

loaka the food that accompanies rice and gives it flavor (Merina)

lôhanaomby head of a zebu; cattle horns

loja groundnut

lokambato stone slab on which offerings to ancestors (such as cooked rice, meat, and rum) are left

mafana hot (adj.); especially strong home-distilled rum (n.)

mahafatoky to be trustworthy

mahatambeloña a type of shrub (lit. "to make living")

malaka to rest, go dormant (as when clove trees stop producing buds several years after efflorescence)

malemy hôngotra lameness ("limp foot/leg")

mamitranjetry a tree with anise-scented leaves (*Vepris lindriana*)

manakalahy an intermediary gait between sleeping water (*rano-mandry*) and racing used by porters

manambany to make low, put down, degrade

manambaotra saina to console; to ease one's mind

manaña aody to have dangerous medicine; to inflict a curse on someone

mañano lagriffe to snap clove buds off their stems

mandeha hôngotra zahe "I am going on foot." (Betsimisaraka)

maromita porter of luggage or cargo in the nineteenth century

masaka cooked

mazoto industrious

mbôla tsara "Still doing well?" (Betsimisaraka greeting)

mila ravin'ahitra "to search for the leaves of weeds" (look for any kind of low-end work to eke out a living)

milela-paladia to kiss the soul of the foot (a gesture of submission)

mitera manambola to show off one has money; to be ostentatious

mitofitofina a gait of porters that employed an uneven step and resulted in passenger fatigue

mora mora easy, soft, mellow (adj. or adv.)

mpamboly/mpambôly cultivator

mpamosavy sorcerer

mpandika liar

mpiasa tany land-worker; peasant

mpiavy inmigrants; newcomers

mpilanja/mpilanjana palanquin carrier

mpisikidy healer or diviner who is able to interpret the meanings of the Sikidy, a form of geomancy based, in Mananara-Nord, on the configuration of piles of seeds

mpitara maso guard ("watching eyes")

mpiteviala forest-clearer; practitioner of *tavy*

nanto species of tree (*Sapotaceae*) used for construction

ny sasany the half

ohotra fern (*Sitcheris flagellaris*)

ôlo maty dead person(s)

ombiasy diviner

orimbato bride price ("laying down a rock" in marriage negotiations between prospective groom and his parents, and the parents of the prospective bride)

pinjy cicada

profo proof

rano mandry sleeping water (a smooth gait used by porters)

ravaraty the razing of dead vegetation (part of the ritual one must do before cultivating ***fanómo***)

rô the food that goes with rice to give it flavor (Betsimisaraka)

sambiremy a species of small river fish

saroño jewelry or clothes given by prospective groom to prospective bride

Savarandrano Mitsinjo Ho Avy "Savarandrano Foresees the Future," the name of a village association involved in community-based forest management

savoka secondary-growth vegetation on a plot where slash-and-burn agriculture has occurred

sazy penalty

silasila razor clam

simbon'tranony razana the robe of the house of ancestors (Betsimisaraka metaphor for the forest)

Sinoa Chinese (Fr. *chinois*)

soherina chrysalis of silkworm moth

sôroko "spoon" made of a soft leaf

sovazy savage (Fr. *sauvage*)

tahôla bone

taimbintaña a species of large, green beetle

tangalamena ("red baton") Betsimisaraka male elder who serves as the spiritual leader of a village and presides over ceremonies

tanindrazaña ancestral land

tantsaha peasant; cultivator

tany soil; land

taoliñy corrugated sheet metal used for roofing (Fr. *tôle*)

tavy shifting, swidden horticulture based on the cultivation of dry hill rice

tenina cogongrass

teza heartwood

Tiako Madagasikara "I love Madagascar," the name of the political party of Marc Ravalomanana

toaka rum

tody finomaña a historical form of punishment (or "trial by ordeal") in which someone accused of sorcery had to drink the poison of a tangena nut

tolongoala a tree with lemon-scented leaves (*Vepris nitida*)

tompony/tômpony property owner

tontolo iainana "the lived-in world"; word invented by conservation planners to signify "the environment"

totroka a plant use to ease stomach ailments

tromba a spirit that possesses a living person

tsaboraha ritual for someone's ancestor(s) in which a zebu is sacrificed and the village feasts on the meat

tsingy limestone formations in north Madagascar

tsiñy a type of wild spirit that inhabits natural features, such as waterfalls or trees, for example

tsy mahalala fomba izy "does not know good manners"

tsy mahay mandeha "does not know how to walk"

tsy mahay miasa "does not know how to work"

tsy mamaky euphemism for "rat" on islet of Nosy Antafaña

tsy mietsika to not move; to be stuck

vaky lôha to be very upset, devastated ("broken-headed")

vazaha stranger, foreigner (typically used for white Europeans or North Americans)

vilompony species of rosewood (formerly used for coffins)

visaka exhausted

voay crocodile

vokatra/vôkatra harvest

vôlamaintiñy "black gold" (in reference to vanilla)

volovava sea urchin

zàna-dàndy (lit. "offspring of silk"), euphemism for the chrysalis of a silkworm moth

Zanahary God

zanatanety "child of the mountains"; peasant

zanatany someone born and raised as a peasant; a local; ("child of the land"; Betsimisaraka)

zavaboary natural resources; the gifts of God to human beings

zaza-vatana polio ("child body")

NOTES

1. Geographies of Borrowed Time

1. The coup d'état, which the opposition labeled an "auto-putsch"—the fate of a rent-seeking, self-interested president, they claimed—was the direct result of an imminent land deal between Daewoo Logistics, a South Korean company, and the government of Madagascar (www.globalwitness.org, accessed 08/19/2004). In November 2008, Daewoo Logistics was on the verge of securing a ninety-nine-year lease for more than 3 million acres of land (an amount representing close to half of the island's arable land) for the production of maize and palm oil to meet Seoul's food demand (Spencer 2008). In exchange for creating jobs, Daewoo was to have been given the lease virtually free of charge. President Marc Ravalomanana's critics in Madagascar accused him of starving the citizenry and pandering to neocolonial interests. One South Korean reporter relished the irony of a report in the *Financial Times* of Britain, erstwhile imperialist power, by a commentator who denounced the Daewoo land deal as "neocolonial" and "rapacious," comparing it to piracy off Somalia (Song et al. 2008; Koehler 2009). The new leadership in Madagascar (still officially unrecognized by the rest of the world at the time of this writing) canceled the contract with Daewoo.

2. Since January 2009, at least US$220 million worth of timber has been looted from protected areas in Madagascar (Randriamalala and Lui 2010). The British photographer Toby Smith documented government officials abetting the rosewood operations during his undercover mission into Madagascar to investigate the rosewood crisis in 2009 (www.shootunit.com, accessed November 2010).

3. The investigative report compiled by the organization Global Witness and the U.S. Environmental Investigation Agency estimates that in 2009 between US$88,000 and US$460,000 worth of rosewood logs were felled daily. This is based on field observations in the Masoala National Park and information about similar illegal activities in the Mananara Biosphere Reserve (Global Witness and the Environmental Investigation Agency 2009).

4. Nashville-based Gibson Guitars was raided in 2009 by U.S. federal agents for possession of endangered Malagasy rosewood. Tipped off by the evidence gathered by Global Witness and the Environmental Investigation Agency, federal agents seized boxes of the rosewood in Gibson's corporate office (Lind 2009).

5. Scientists estimate the number of vascular plant species to be 8,500, and the total number of plant species to be 10,000 to 12,000 (Dorr, Barnett, and Rakotozafy 1989:238; Feeley-Harnik 2001:36).

6. Myers further defines a biodiversity hot spot as a region that contains at least 70 percent of its primary vegetation, of which at least 1,500 species of vascular plants are endemic (Myers et al. 2000).

7. Pierre Boiteau (1958:225) writes that in 1954, six years before Madagascar's independence, the French state had expropriated around 10 million hectares of primary forest in Madagascar, about 9 percent of which it classified as integral reserves (*réserves intégrales*), strictly off-limits for any purpose other than authorized scientific research. Michael Williams (2003) confirms: "Of the 58 million ha of the island, by 1920 about 20 percent was said to remain in primary forest. By 1949 it was 8.6 percent; 10.3 percent was in *savoka,* second-growth bracken fern with bamboos; and the rest was in grass and savanna" (343).

8. In his review of data and studies concerning rates of deforestation, Christian Kull (2004) writes that to date, Green and Sussman's 1990 study is the best-documented source on the matter. Green and Sussman's study estimates a 66 percent loss of primary forest coverage. At 1980s' rates of deforestation, the eastern rain forest would disappear by 2025 apart from stands on the steepest slopes. Yet it should be noted that the rates, if not the fact, of forest loss remain contested (Kull 2004:161–164; International Monetary Fund 2007:27). The evolution of forest to grassland varies depending on "soils, topography, surrounding land use patterns, and the frequency and length of cultivation," and can take anywhere from over a century to fifteen to twenty years (Kull 2004:160).

9. In seeking to correct the bias of the colonial narrative, contemporary scholars of Madagascar have often romanticized peasant land use practices or blamed land degradation on unfair policies (Pollini 2010). Geographer Jacques Pollini (2010:711) asserts that like the colonial narrative that vilifies *tavy,* left-leaning counternarratives also misrepresent the history of environmental change. His argument recalls the critique of political ecology by Andrew P. Vayda and Bradley B. Walters (1999), who regret the attenuation of "ecology" in the political ecology literature, which they attribute to the proclivity of self-identified political ecologists to interpret all environmental change as an effect of unequal power relations and resource distribution. This analytical slant, they argue, minimizes the effects of natural forces on landscapes and social transformations.

10. Dimpault, "Mission Pegourier: Forêts," 1928; FR ANOM GGM MAD 3D c. 13.

11. Rosaleen Duffy (2008:332–333) writes that over "the period 1990–2000 visitor arrivals to Madagascar increased by 202 percent, with 53,000 visitors in 1990 and 160,000 visitors in 2000. . . . World Tourism Organisation data also shows that Madagascar obtained US$105 million in tourism receipts in 2005 (UNWTO 2006)."

12. UNESCO, established in 1945, stands for the United Nations Educational, Scientific, and Cultural Organization.

13. In Mananara-Nord, the Merina residents I knew resented but tolerated the name *hôva,* explaining that Betsimisaraka called them such due to their ignorance of the real meaning of the term. Merina society was historically divided as follows: *andriana* represented the noble class; *hôva* represented free people; and *andevo* represented slaves. Merina today of middle or upper classes take pride in their right to claim noble heritage; thus the term *hôva* may be degrading.

2. Overland on Foot, Aloft

1. Smith was hired in 2009 by the organizations Global Witness and the Environmental Investigation Agency, in collaboration with United States federal authorities, to collect evidence of the illegal rosewood operations under way in northeast Madagascar. Pretending to be bird-watchers and then buyers of rosewood, ebony, and palisander, Smith and his companions traveled to the logger encampments and spoke with the crews of men who felled timber and conveyed it downriver to the coast. They also spoke to some of the Timber Barons and corrupt officials that abetted the rosewood trade. Although he does not say so explicitly so as not to implicate or endanger the men, Smith's team of three to five investigators depended on resident guides to safely navigate the rain forest and the logger encampments (Smith, personal communication, 7/13/2010).

3. Land and Languor

The second epigraph is from a paper presented at the Congrès International et Intercolonial de la Société Indigène, Paris, 5–10, October 1931, BIB AOM B6620; my translation. An earlier version of this chapter appears in Genese Sodikoff, 2004, "Land and Languor: Ethical Imaginations of Work and Forest in Northeast Madagascar," *History and Anthropology* 15(4): 367–398.

1. In 1900, the "grand forest" stretched from the southern limit of Tamatave province to the Anove River of the Mananara region, starting about 40 kilometers inland and parallel to the coast. Jeannelle, "Rapport de Tournée, 5 mai," 1900; FR ANOM GGM MAD 6D(9), c. 7. None of the early annual reports about Mananara-Nord offers a total surface area size for forest, only the sizes of individual concessions. Exploitable concessions granted to French entrepreneurs between 1901 and 1903 alone comprised a total of 121,690 hectares; Chef du District de Mananara, "Etat des Exploitations Forestières en 1er janvier, Statistiques," 1903; FR ANOM GGM MAD 2D, c. 153PO.

2. Fiangonan'i Jesoa Kristy eto Madagasikara (Church of Jesus Christ in Madagascar).

3. Quartz crystal is also known as *vato mazava,* "clear rock," in Merina.

4. The spiritually sanctioned social hierarchy that underpinned Merina society (see Berg 1995; Bloch 1986) appeared to complement the ideology of colonial bureaucracy. Merina subjects of the French empire accepted Europeans' entrenched conceptions of other Malagasy groups as they were assimilated into the colonial bureaucracy as work site overseers and low-level functionaries.

5. Stefani, "Rapport Economique, District de Mananara," 1914; FR ANOM GGM MAD 2D, c. 140.

6. Chef du District de Mananara, "Rapport Economique 2e Semestre, 1906," FR ANOM GGM MAD 2D, c. 140.

7. Jeannelle, "Rapport de Tournée, 5 mai, 1900," FR ANOM GGM MAD 6D(9), c. 7. Jeannelle's report is addressed to the governor general, Joseph Simon Gallieni (ruled 1897–1905). Subsequent quotations of Jeannelle derive from the same source. During this period, Mananara was a district of Maroantsetra province, but by 1932 it was redistricted into Toamasina (Tamatave) province, where it remains.

8. Chef du District de Mananara, "Rapport Economique, 1er Semestre, 20 juin," 1906; FR ANOM GGM MAD 2D, c. 140.

9. Chef du District de Mananara, "Rapport de Province, 30 septembre," 1910; FR ANOM GGM MAD 2D, c. 140.

10. Chef du District de Mananara, "Rapport Economique," 1912; FR ANOM GGM MAD 2D, c. 140.

11. Stefani, "Rapport Economique," 1913; FR ANOM GGM MAD 2D, c. 140.

12. Ibid.

13. Peter Sahlins (1994) discusses the problem of "disorderly" peasant practices in seventeenth-century France. At this time the monarchy undertook its first comprehensive steps to administer forest lands and impose order.

14. Stefani, "Journal du District de Mananara, 2e Semestre," 1914; FR ANOM GGM MAD 2D, c. 140.

15. Chef du District de Mananara, "Etat des Exploitations Forestières en 1er janvier, Statistiques," 1903; FR ANOM GGM MAD 2D, c. 153PO.

16. Lagriffoul, "Lettre à Gouverneur-Général, No 769, 17 juin," 1902; FR ANOM GGM MAD 6D(5), c. 8.

17. Lagriffoul, "Lettre à Gouverneur-Général, No 1160, 8 octobre," 1902; FR ANOM GGM MAD 6D(5), c. 8.

18. Chef de District de Mananara, "Rapport Politique et Administratif, 9 février," 1911; FR ANOM GGM MAD 2D, c. 140.

19. In 1933, for example, Governor-General Léon Cayla complains: "The natives of the canton of Mananara are not ignorant of the binding law or to the sanctions to which they expose themselves in violating the law. But their natural laziness leads them despite everything to practice *tavy*." Cayla, "Menées de Railaimongo; Incident de Mananara, No 356, 21 février," 1933; FR ANOM GGM MAD 6D(2), c. 108. My translation.

20. Dumont, "Lettre Confidentielle au Chef de la Region de Tamatave, No 2634, 22 décembre," 1932; FR ANOM GGM MAD 6D(2), c. 108.

21. The *fokontany* is the smallest territorial division that represents the state. The *president-pokontany* (president of the *fokontany*) is otherwise called the resident of the local security committee (CLS). In contrast to the *fokontany*, the *fokonolona* represents the customary authority of ancestors and the spiritual chief, or *tangalamena* (Locatelli 2000).

22. The dates are uncertain. The *tangalamena* claimed that the Mananara-Nord post office was completed in 1920.

23. Elsewhere in Madagascar, however, corrupt work site bosses allowed private entrepreneurs to occasionally obtain forced laborers for their private concessions. This practice was certainly in effect by the 1920s, if not before (Stratton 1964).

24. The International Labor Organization's Convention against Forced and Compulsory Labor, which France refused to sign until 1937, put the French on the defensive in international forums regarding SMOTIG and analogous colonial labor regimes.

25. Along with his followers, the Betsileo-born Ralaimongo, an anti-colonialist intellectual who disseminated communist literature, inspired a popular protest against the prohibition of *tavy* in Mananara-Nord in the early 1930s. For several years, district administrative reports focused on the "Ralaimongo Affair" and interrogations of his accomplices in Mananara-Nord. See, for example, Boiteau, "Rapport sur les collectes d'argent dans le district de Mananara, 30 janvier," 1933; FR

ANOM GGM MAD 6D(2), c. 108; Cayla, "Lettre Confidentielle au Ministre des Colonies, Direction des Affaires Politiques, No 356, 21 février," 1933; FR ANOM GGM MAD 6D(2), c. 108.

26. Chef du District de Mananara, "Etat des Exploitations Forestières en 1er janvier, Statistiques," 1903; FR ANOM GGM MAD 2D, c. 153PO.

27. Chef du District de Mananara, "Rapport Economique 1er Semestre," 1906; FR ANOM GGM MAD 2D, c. 140.

28. Ibid.

29. Royet, Chef du District de Mananara, "Rapport Economique, L'adjoint des affaires civiles, 31 janvier," 1901; FR ANOM GGM MAD 2D, c. 140.

30. Chef du District de Mananara, "Rapport Economique 2e Semestre," 1906; FR ANOM GGM MAD 2D, c. 140.

31. Fillastre, "Rapport, No 63, 19 avril," 1910; FR ANOM GGM MAD 2D, c. 140.

32. Chef du District de Mananara, "Rapport Politique et Administratif, janvier-février," 1911; FR ANOM GGM MAD 2D, c. 140.

33. Chef du District de Mananara, "Rapport Economique 2e Semestre," 1906; FR ANOM GGM MAD 2D, c. 140.

34. Cooper writes that elsewhere in Africa, for example in Kenya, colonial officials were opposed to the presence of "casual laborers," who escaped "the discipline of regular work." Even though no evidence proved that casual labor was less efficient than regular labor, "casual laborers exercised too much choice in the labor process" (Cooper 1996:236). See also Cooper 1992.

35. Chef du District de Mananara, "Rapport Politique et Administratif, janvier-février," 1911; FR ANOM GGM MAD 2D, c. 140.

4. Toward a New Nature

Sections of this chapter appear in Genese Sodikoff, 2005, "Forced and Forest Labor Regimes in Colonial Madagascar, 1926–1936." *Ethnohistory* 52(2): 407–435.

1. See annual reports of the Service de l'Agriculture and the Service Forestier, FR ANOM GGM MAD 5 D (18), c. 1–20.

2. Griess, "Note Sur le Service des Forêts a Madagascar," 1923; FR ANOM GGM MAD 5D(18), c. 9. My translation.

3. Ibid., 3.

4. Ibid., 14.

5. Dimpault, "Mission Pegourier: Forêts," 1928; FR ANOM GGM MAD 3D, c. 13.

6. Laurence Dorr (1997:214–218) describes Jean Henri Humbert (1887–1967) as a "taxonomist, phytogeographer, conservationist, and one of the foremost explorers of Madagascar."

7. For a summary of legislative acts protecting forested *domaines* through 1932, see "Inspection des Provinces Répertoire Analytique des Principaux Textes en vigueur Arrêté au 31 avril 1926; Supplement au J.O," 1932; BIB AOM 41903.

8. Poirier, "Lettre Confidentielle à M. Chataux, Chef du District d'Ivohibe, Rapport No. 103, janvier," 1928; FR ANOM GGM MAD 3D, c. 114.

9. Analamazoatra is current site of the project office of the Andasibe-Mantadia Protected Area Complex.

10. Lavauden, "Rapport général sur le fonctionnement du Service forestier," 1930; FR ANOM GGM MAD 5D(18), c. 15: 11.

11. Poirier, "Lettre Confidentielle à M. Chataux, Chef du District d'Ivohibe, Rapport No. 103, janvier, 1928"; FR ANOM GGM MAD 3D, c. 114.

12. Calculations in Dimpault's 1928 report on the forest service show that the salary for a European principal forest guard was 25,500 F/month, while a Malagasy principal guard earned 3,800 F/month.

13. Lavauden, "Rapport général sur le fonctionnement du Service forestier," 1930; FR ANOM GGM MAD 5D(18), c. 15.

14. From a confidential letter from the Chef de la Province de Moramanga to Governor General Olivier, "Objet: A.S des mesures à prendre contre la déforestation, No 277, 3 juin," 1929; FR ANOM GGM MAD 5 D (18), c. 15.

15. Lavauden, "Rapport général sur le fonctionnement du Service forestier," 1930; FR ANOM GGM MAD 5D(18), c. 15: 17.

16. In the confidential letter from the Chef de la Province de Moramanga to the governor-general (June 3, 1929), the author notes the "insufficiency of forestry personnel and the passivity of indigenous surveillance agents in this matter"; FR ANOM GGM MAD 5 D(18), c. 15. My translation.

17. Dimpault, "Mission Pegourier: Forêts," 1928; FR ANOM GGM MAD 3D, c. 13. My translation.

18. Lavauden, "Rapport général sur le fonctionnement du Service forestier," 1930; FR ANOM GGM MAD 5D(18), c. 15. My translation.

19. Ibid.

20. Marius Moutet, Jore, et al., incomplete notes on the "Service des Forêts" and "SMOTIG," 1936; FR ANOM GGM MAD 3A(2), c. 15.

21. Direction de la Production du Sol, "Rapport Annuel Exposant le Fonctionnement et l'Activité de la Direction de la Production du Sol au Cours de l'Année," 1945; FR ANOM GGM MAD 5D(18), c. 40.

22. Direction de la Production du Sol, "Rapport Annuel: Fonctionnement du Bureau de Conservation des Sols de Madagascar et Dependances au Cours de l'Année," 1950; FR ANOM GGM MAD 5D(18), c. 43.

23. A 1951 forest service report cited by A. Keiner, Inspecteur Principal des Eaux et Forêts, in "Esquisse Forestière de la Province de Tamatave, Aperçu de quelques Problèmes Forestiers et de Conservation des Sols en 1954"; *Bulletin de Madagascar* 133 (1957); BIB AOM B6406.

24. Ibid., 25.

25. In Madagascar today there are eight national trade union centers, including Fivondronamben'ny Mpiaga Malagasy (FMM) with a total membership of 30,000; Union Syndicats Autonomes de Madagascar (USAM) with a total membership of 2,706; Sendika Krtianina Malagasi (SEKRIMA) with a total membership of 5,000; Fédération du Travailleurs Malagasy Revolutionnaires (FISEMA), Cartel National des Organisations Syndicale de Madagascar (CARNOSYMA), Fédération des Syndicats de Travailleurs de Madagascar (FISEMARE), with a total membership of 60,000; Sendika Revolisakonera Malagasy (SEREMA); and Union Syndicale des Travailleurs et Paysans Malagaches (USTPM). Statistics from the Ministry of Labour indicate that less than 10 percent of all workers are unionized. This is largely due to the fact that 80 percent of the workforce is engaged in subsistence agriculture (http://www.icftuafro.org/public/ProfilesEN.pdf).

6. How the Dead Matter

1. At the time, Club Med had recently abandoned its plans to construct a resort in the north due to conflict with the state over this issue of "cultural corruption," suggesting a notion of cultural containment on the part of Malagasy Ministry representatives. These officials feared, according to my informant, the transformation of Madagascar into a place like the Caribbean.

2. Feeley-Harnik notes that among the Sakalava, shrubs called *matambelona,* or "dead-living," were planted with those called *hasina,* or "generative" (1991:169, 184), whereas in the village of Varary in the Mananara-Nord region, Betsimisaraka pronounced the shrubby tree *mahatambeloña.*

3. James Sibree (1870:58) records his sight of the tangena during his travels in Madagascar in the late nineteenth century: "The tangéna is about the size of a large apple-tree, and could it be naturalized in England would make a beautiful addition to our ornamental plantations. . . . The poison is procured from the nut, and was, until a very recent period, used with fatal effect for the trial of accused persons, and caused the death of many thousands of innocent people during the reign of Queen Rànavàlona."

4. M. Jeannelle, "Rapport de Tournée, 5 mai," 1900; FR ANOM GGM MAD 6D(9), c. 7.

7. Cooked Rice Wages

1. An important way for living kin to pay tribute to their deceased parents or other relatives and to improve the chances of reaping an ample rice crop at harvest was by sacrificing a zebu, which for most households constituted a major expenditure. The public event was called *tsaboraha* ("doing something," or "caring for something"). *Tsaboraha* is linked to the verb *mitsabo,* which among Sakalava signified the work of "growing/attending/nursing crops" (Feeley-Harnik 1991:192). The entire village attended the feast. The *tsaboraha* signified the act of "giving the ancestors their share" (*rasahariaña*) of the bounty enjoyed by the living. It was the fulfillment of a promise by the living to the dead: if a person prays to the ancestors for something that comes to pass, such as the birth of a healthy child after many miscarriages, he or she must give a share of wealth back in the form of zebu as recompense (*tsika fara*). Zebus are extremely expensive for rural farmers, each one costing between approximately 700,000 and 1,000,000 FMG in 2001 (US$112–160). In addition, the host of the *tsaboraha* must meet the cost of furnishing rice and drink for the village during the event.

Epilogue

1. American ecologist Warden Clyde Allee (1931) described the effect where in declining or small animal populations, a variety of different processes may reduce an individual's evolutionary fitness. As population levels grow, this effect usually disappears.

BIBLIOGRAPHY

Colonial Archival Documents

FR ANO Archives National Outre-Mer, Aix-en-Provence, France (formerly referenced as CAOM, Centre des Archives d'Outre-Mer)

GGM Gouvernement Générale de Madagascar 1841–1960

This collection is divided into four series:

Série A. Actes officiels 1896–1959

Série B. Correspondance générale et registres divers 1841–1960

Série C. Personnel européen XIXe–XXe siècles

Série D. Affaires politiques et administration générale 1895–1959

Série D is subdivided into six subseries [*sous-séries*]

1D. Correspondance générale de la direction des affaires poliques

2D. Rapports périodiques des circonscriptions administratives

3D. Missions d'inspection

4D. Rapports généraux sur la situation de Madagascar

5D. Rapports annuels des services

6D. Dossiers divers de la Direction des affaires politiques

MAD Madagascar (c.=carton)

BIB AOM Bibliothèque des Archives d'Outre-Mer [collections housed within FR ANOM]

Published Sources, Digital Sources, and Manuscripts

Abinal, R. P., and V. Malzac

2000 [1888] Dictionnaire Malgache-Français. Analamahitsy, Antananarivo, Madagascar: Editions Ambozontany.

Adas, Michael

1986 From Footdragging to Flight: The Evasive History of Peasant Avoidance Protest in South and South-east Asia. Journal of Peasant Studies 13(2): 64–86.

Afrol News

2005 Greater Pressure on Madagascar's Media. http: www.afrol.com/articles /15826, accessed June 15, 2005.

Aguiar, João

2007 Capital and Nature: An Interview with Paul Burkett. MR Zine, April 24.

http://www.monthlyreview.org/mrzine/aguiar240407.html, accessed December 5, 2010.
Agrawal, Arun
2005 Environmentality: Technologies of Government and the Making of Subjects. Durham, N.C.: Duke University Press.
Agrawal, Arun, and Clark Gibson, eds.
2001 Communities and the Environment: Ethnicity, Gender, and the State in Community-Based Conservation. New Brunswick, N.J.: Rutgers University Press.
Allee, Warden C.
1931 Animal Aggregations: A Study in General Sociology. Chicago: University of Chicago Press.
Allen, Philip M.
1995 Madagascar: Conflicts of Authority in the Great Island. Boulder, Colo.: Westview Press.
Althabe, Gérard
2002 Oppression et Libération dans l'Imaginaire: Les communautés villageoises de la côte orientale de Madagascar. Paris: La Découverte.
Anderson, David
1984 Depression, Dust Bowl, Demography, and Drought: The Colonial State and Soil Conservation in East Africa during the 1930s. African Affairs 83(332): 321–343.
2002 Eroding the Commons: The Politics of Ecology in Baringo, Kenya 1890–1963. Oxford: James Currey.
Anderson, David G., and Eeva Berglund, eds.
2003 Ethnographies of Conservation: Environmentalism and the Distribution of Privilege. Oxford: Berghahn.
Anderson, David, and Richard Grove, eds.
1987 Conservation in Africa: People, Policies and Practice. Cambridge: Cambridge University Press.
Andrew, David, Becca Blond, Tom Parkinson, and Aaron Anderson
2008 Lonely Planet Madagascar and Comoros. 6th edition. Oakland, Calif.: Lonely Planet Publications.
Angier, Natalie
2009 New Creatures in an Age of Extinctions. New York Times, July 26: 1–3.
Applebaum, Herbert, ed.
1984 Work in Market and Industrial Societies. New York: State University of New York Press.
Arbousset, Francis
1950 Le Fokonolona à Madagascar. Paris: Domat-Montchrestien.
Archer, Robert
N.d. Madagascar Depuis 1972. Paris: L'Harmattan.
Argyrou, Vassos
2005 The Logic of Environmentalism: Anthropology, Ecology, and Postcoloniality. New York: Berghahn Books.
Arnold, David
1996 The Problem of Nature: Environment, Culture and European Expansion. Oxford: Blackwell.

Atkins, Keletso E.
1993 The Moon Is Dead! Give Us Our Money! The Cultural Origins of an African Work Ethic, Natal, South Africa, 1843–1900. Portsmouth, N.H.: Heinemann.
BBC News
2005 Madagascar's Conservation Conundrum. April 12. http://news.bbc.co.uk/2/hi/africa/4433817.stm, accessed March 2, 2009.
Baron, Richard
1892 Twelve Hundred Miles in a Palanquin. *In* The Antananarivo Annual, no. 16. J. Sibree and R. Baron, eds. Pp. 434–458. Antananarivo, Madagascar: LMS Press.
Barrett, Christopher B.
1994 Understanding Uneven Agricultural Liberalisation in Madagascar. Journal of Modern African Studies 32(3): 449–476.
Batisse, Michel
1997 Biosphere Reserves: A Challenge for Biodiversity Conservation and Regional Development. Environment 39(5): 7–33.
Bearak, Berry
2010 Scarce Madagascar Rosewood Plundered. New York Times, June 13. P. A1.
Beattie, Andrew, Paul Ehrlich, and Christine Turnbull
2001 Wild Solutions: How Biodiversity Is Money in the Bank. New Haven, Conn.: Yale University Press.
Beaujard, Philippe
1998 Dictionnaire Malgache-Français: Dialecte Tañala, Sud-Est de Madagascar. Paris: L'Harmattan.
Beidelman, T. O.
1980 The Moral Imagination of the Kaguru: Some Thoughts on Tricksters, Translation and Comparative Analysis. American Ethnologist 7(1): 27–42.
Beinart, William
1984 Soil Erosion, Conservationism and Ideas about Development: A Southern African Exploration, 1900–1960. Journal of Southern African Studies 11: 52–83.
1989 Introduction: The Politics of Colonial Conservation. Journal of Southern African Studies 15(2): 143–162.
1996 Environmental Destruction in Southern Africa: Soil Erosion, Animals, and Pastures over the Longer Term. *In* Time-Scales and Environmental Change. Thackwray S. Driver and Graham Chapman, eds. Pp. 149–168. London: Routledge.
Bellah, R. N., R. Madsen, W. M. Sullivan, A. Swidler, and S. M. Tipton
1985 Habits of the Heart: Individualism and Commitment in American Life. Berkeley: University of California Press.
Benería, Lourdes, and Gita Sen
1981 Accumulation, Reproduction, and "Women's Role in Economic Development": Boserup Revisited. Signs 7(2): 279–298.
Benton, Ted, ed.
1996 The Greening of Marxism. New York: Guildford.
Berg, Gerald M.
1981 Riziculture and the Founding of Monarchy in Imerina. Journal of African History 22(3): 289–308.

1995 Writing Ideology: Ranavalona, the Ancestral Bureaucrat. History in Africa 22: 73–92.
Bergeret, Anne
1993 Discours et Politiques Forestières Coloniales en Afrique et à Madagascar. Revue Française d'Histoire d'Outre-Mer 79(298): 23–47.
Bernal, Victoria
1991 Cultivating Workers: Peasants and Capitalism in a Sudanese Village. New York: Columbia University Press.
Berner, P.
1993 Conservation-Based Forest Management Development Strategies to Reduce Pressure on the Analamazaotra-Mantadia Complex, Madagascar. Antananarivo, Madagascar: VITA and Tropical Forest Management Trust.
Bernstein, Basil
1971 Class, Codes, and Control, vol. 1. London: Routledge and Kegan Paul.
Berry, Sara
1985 Fathers Work for Their Sons: Accumulation, Mobility, and Class Formation in an Extended Yorùbá Community. Berkeley: University of California Press.
2002 Debating the Land Question in Africa. Comparative Studies in Society and History 44(4): 638–668.
Bertrand, Alain
1999 La Gestion Contractuelle, Pluraliste et Subsidiaire des Ressources Renouvelables à Madagascar (1994–1998). African Studies Quarterly 3(2): 75–81.
Bertrand, Alain, and M. Sourdat
1998 Feux et Déforestation à Madagascar, Revues Bibliographiques. Antananarivo, Madagascar: CIRAD/ORSTOM/CITE.
Bertrand, Alain, Pierre Montagne, and Alain Karsenty, eds.
2006 Forêts Tropicales et Mondialisation: Les mutations des politiques forestières en Afrique francophone et à Madagascar. Paris: L'Harmattan.
Biersack, Aletta
2006 Reimagining Political Ecology: Culture/Power/History/Nature. *In* Re-Imagining Political Ecology. Aletta Biersack and James Greenberg, eds. Pp. 3–39. Durham, N.C.: Duke University Press.
Biersack, Aletta, and James Greenberg, eds.
2006 Re-imagining Political Ecology. Durham, N.C.: Duke University Press.
Black, Richard
2010 Million-Dollar Beds Fuel Madagascar Timber Crisis. BBC News, Science and Environment. October 26. http://www.bbc.co.uk/news/science-environment-11626412, accessed April 10, 2011.
Blaikie, Piers
1985 The Political Economy of Soil Erosion in Developing Countries. London: Methuen.
2001 Social Nature and Environmental Policy in the South: Views from Verandah and Veld. *In* Social Nature: Theory, Practice, and Politics. Noel Castree and Bruce Braun, eds. Pp. 133–150. Oxford: Blackwell.
Blaikie, P. M., and H. C. Brookfield
1987 Land Degradation and Society. London: Methuen.

Blench, Robert
2006 The Austronesians in Madagascar and on the East African Coast: Surveying the Linguistic Evidence for Domestic and Translocated Animals. Paper given at the International Conference on Austronesian Languages, Puerto Princesa, Palawan, January 17–20. Draft.
Bloch, Maurice
1971 Placing the Dead: Tombs, Ancestral Villages, and Kinship Organization in Madagascar. London: Seminar.
1980 Modes of Production and Slavery in Madagascar. *In* Asian and African Systems of Slavery. James L. Watson, ed. Pp. 100–134. Oxford: Oxford University Press.
1986 From Blessing to Violence: History and Ideology in the Circumcision Ritual of the Merina of Madagascar. Cambridge: Cambridge University Press.
1995 People into Places: Zafimaniry Concepts of Clarity. *In* The Anthropology of Landscape: Perspectives on Place and Space. Eric Hirsch and Michael O'Hanlon, eds. Pp. 63–77. New York: Oxford University Press.
Boiteau, Pierre
1958 Contribution à l'Histoire de la Nation Malgache. Paris: Editions Sociales.
Bouche, Denise
1991 Histoire de la Colonisation Française. Tome Second: Flux et reflux (1815–1962). Paris: Fayard.
Bourdieu, Pierre
1984 Distinction. London: Routledge and Kegan Paul.
Bourton, Jody
2009 Lemurs Butchered in Madagascar. BBC Earth News (British Broadcasting Corporation). August 20.
Boussenot, Georges
1925 La Presse Coloniale Illustrée: Madagascar Industriel. Paris: La Presse Coloniale Illustrée.
Bowman, D. M. J. S.
2001 Future Eating and Country Keeping: What Role Has Environmental History in the Management of Biodiversity? Journal of Biogeography 28: 549–564.
Brand, Jürg
2000 Réserve de Biosphère Mananara-Nord, Plan de Gestion 2002–2006. Draft Report, February. Antananarivo, Madagascar: Development Environment Consult.
Brand, J., and J. L. Pfund
1998 Site- and Watershed-Level Assessment of Nutrient Dynamics under Shifting Cultivation in Eastern Madagascar. Agriculture, Ecosystems and Environment 71(1/3): 169–183.
Brandon, Katrina, and Michael Wells
1992 Planning for People and Parks: Design Dilemmas. World Development 20(4): 557–570.
Brantlinger, Patrick
2003 Dark Vanishings: Discourse on the Extinction of Primitive Races, 1800–1930. Ithaca, N.Y.: Cornell University Press.

Breidenbach, Joana, and Pál Nyíri
2007 "Our Common Heritage": New Tourist Nations, Post-"Socialist" Pedagogy, and the Globalization of Nature. Current Anthropology 48(2): 322–330.
Bressers, Hans T. A., and Walter A. Rosenbaum, eds.
2003 Achieving Sustainable Development: The Challenge of Governance across Social Scales. Westport, Conn.: Praeger.
Brinkmann, Svend
2004 The Topography of Moral Ecology. Theory and Psychology 14(1): 57–80.
Brockington, Dan
2002 Fortress Conservation: The Preservation of the Mkomazi Game Reserve, Tanzania. Bloomington: Indiana University Press.
Brosius, J. Peter
1999 Green Dots, Pink Hearts: Displacing Politics from the Malasian Rainforest. American Anthropologist 101(1): 36–57.
Brosius, J. Peter, Anna Lowenhaupt Tsing, and Charles Zerner, eds.
2005 Communities and Conservation: Histories and Politics of Community-Based Natural Resource Management. Walnut Creek, Calif.: AltaMira Press.
Brown, Carolyn A.
1988 The Dialectics of Colonial Labour Control: Class Struggles in the Nigerian Coal Industry, 1914–1949. Journal of Asian and African Studies 23(1–2): 32–59.
Brown, Margaret
1999 Authority Relations and Trust: Social Cohesion on the Eastern Masoala Peninsula, Madagascar. PhD Dissertation, Washington University, Saint Louis.
2004 Reclaiming Lost Ancestors and Acknowledging Slave Descent: Insights from Madagascar. Comparative Studies in Society and History 46(3): 616–645.
Brown, Michael F.
2004 Heritage as Property. *In* Property in Question: Value Transformation in the Global Economy. Katherine Verdery and Caroline Humphrey, eds. Pp. 49–68. New York: Berg.
Bryant, Raymond L.
1996 Romancing Colonial Forestry: The Discourse of "Forestry as Progress" in British Burma. Geographical Journal 162(2): 169–178.
2005 Nongovernmental Organizations in Environmental Struggles: Politics and the Making of Moral Capital in the Philippines. New Haven, Conn.: Yale University Press.
Bryant, R. L., and S. Bailey
1997 Third World Political Ecology. London: Routledge.
Bundy, Colin
1979 The Rise and Fall of the South African Peasantry. London: Heinemann.
Burkett, Paul
1999 Marx and Nature: Red and Green Perspective. New York: St. Martin's Press.
Burney, David A.
1997 Theories and Facts Regarding Holocene Environmental Change before and after Human Colonization. *In* Natural Change and Human Impact in Madagascar. Steven M. Goodman and Bruce D. Patterson, eds., 75–89. Washington, D.C.: Smithsonian Institution Press.

Burney, David A., Lida Pigott Burney, Laurie R. Godfrey, William L. Jungers, Steven M. Goodman, Henry T. Wright, and A. J. Timothy Jull.
2004 A Chronology for Late Prehistoric Madagascar. Journal of Human Evolution 47: 25–63.
Burney, David A., and Ramilisonina
1998 The Kilopilopitsofy, Kidoky, and Bokyboky: Accounts of Strange Animals from Belo-sur-mer, Madagascar, and the Megafaunal "Extinction Window." American Anthropologist 100(4): 957–966.
Caldwell, Lynton Keith
1996 International Environmental Policy: From the Twentieth to the Twenty-First Century. Durham, N.C.: Duke University Press.
Calvino, Italo
1981 If on a Winter's Night a Traveler. Translated from the Italian by William Weaver. San Diego, Calif.: Harcourt Brace Jovanovich.
Campbell, Gwyn
1988 Slavery and Fanompoana: The Structure of Forced Labour in Imerina (Madagascar), 1790–1861. Journal of African History 29: 463–486.
2005 An Economic History of Imperial Madagascar, 1750–1895: The Rise and Fall of an Island Empire. Cambridge: Cambridge University Press.
Capital Institute
2010 Can Nature Be Monetized? A Capital Institute Conversation. Electronic text. http://www.capitalinstitute.org/forum/ecosystem/can-nature-be-monetized-capital-institute-conversation, accessed August 24, 2010.
Carapico, Sheila
2000 NGOs, INGOs, GO-NGOs and DO-NGOs: Making Sense of Non-Governmental Organizations. Middle East Report 214, 30(1): 12–15.
Carrier, James G.
2001 Social Aspects of Abstraction. Social Anthropology 9(3): 243–256.
2004 Environmental Conservation and Institutional Environments in Jamaica. *In* Confronting Environments: Local Understanding in a Globalizing World. James Carrier, ed. Pp. 119–141. Walnut Creek, Calif.: AltaMira Press.
Carrier, James G., ed.
2004 Confronting Environments: Local Understanding in a Globalizing World. Walnut Creek, Calif.: AltaMira Press.
Castree, Noel
1996/1997 Invisible Leviathan: Speculations on Marx, Spivak, and the Question of Value. Rethinking Marxism 9(2): 45–78.
2001 Commodity Fetishism, Geographical Imaginations and Imaginative Geographies. Environment and Planning A 33(9): 1519–1525.
2008 Neoliberalising Nature: The Logics of Deregulation and Reregulation. Environment and Planning A 40(1): 131–152.
2009 The Spatio-temporality of Capitalism. Time and Society 18: 26–61.
Cayla, Léon
1931 Madagascar et Dépendances: Notice Géographique, Politique et Economique. *In* Le Livre d'Or de L'Exposition Coloniale Internationale de Paris. Pp. 101–105. Paris: Librairie Ancienne Honoré Champion.

Ceballos-Lascuráin, H.
1996 Tourism, Ecotourism and Protected Areas: The State of Nature-Based Tourism around the World and Guidelines for Its Development. Gland, Switzerland: IUCN.
Chakrabarty, Dipesh
2008 The Climate of History. Critical Inquiry 35: 197–222.
Chambers, Robert
1983 Rural Development: Putting the Last First. London: Longman.
Charte de l'Environnement
1990 Loi No 90–033 du 21 Décembre 1990. Journal official de la République Démocratique de Madagascar, 2540–2589. Antananarivo, Madagascar: Foi et Justice.
Chauveau, Etienne, and Thierry Guineberteau
2003 Confrontations Sociétales et Scalaires: Théories et pratiques de la conservation de la nature dans la Réserve de Biosphère de Mananara-Nord (Madagascar). Technical Report. Nantes: Université de Nantes.
Christie, Iain T., and D. Elizabeth Crompton
2003 Republic of Madagascar: Tourism Sector Study. Africa Region Working Paper Series, 63(E). Washington, D.C.: World Bank.
Cohen, David William
1994 The Combing of History. Chicago: University of Chicago Press.
Cohen, Robin
1976 Hidden Forms of Labour Protest in Africa. Paper Presented at Conference on Inequality in Africa, Joint Committee on African Studies, Social Science Research Council, Mt. Kisco, New York, October.
Colding, Johan, and Carl Folke
2001 Social Taboos: "Invisible" Systems of Local Resource Management and Biological Conservation. Ecological Applications 11(2): 584–600.
Cole, Jennifer
2001 Forget Colonialism? Sacrifice and the Art of Memory in Madagascar. Berkeley: University of California Press.
2003 Narratives and Moral Projects: Generational Memories of the Malagasy 1947 Rebellion. Ethos 13(2): 95–126.
Colin, E.
1898 The End of an Observatory. The Observatory 21(269): 305–308.
Collins, John F.
2008 "But What If I Should Need to Defecate in Your Neighborhood, Madame?" Empire, Redemption and the "Tradition of the Oppressed" in a Brazilian Historical Center. Cultural Anthropology 23 (2008): 279–328.
2011 Culture, Content, and the Enclosure of Human Being: UNESCO's "Intangible" Heritage in the New Millennium. Radical History Review 109: 120–136.
Colom, Jacques
1992 La Protection de L'Environnement à Madagascar, à Maurice, aux Comores et aux Seychelles. Annuaire des Pays de l'Océan Indien, 11 (1986–1989): 65–88.
Comaroff, Jean, and John Comaroff
2001 Naturing the Nation: Aliens, Apocalypse and the Postcolonial State. Journal of Southern African Studies 27(3): 627–651.

Comaroff, John, and Jean Comaroff
1987 The Madman and the Migrant: Work and Labor in the Historical Consciousness of a South African People. American Ethnologist 14: 191–209.
Comte, Jean
1963 Les Communes Malgaches. Editions de la Librairie de Madagascar. Antananarivo, Madagascar: Imprimerie Protestante Imarivolanitra.
Cooke, A., J. R. E. Lutjeharms, and P. Vasseur
2004 Marine and Coastal Ecosystems. *In* The Natural History of Madagascar. Steven M. Goodman and Jonathan P. Benstead, eds. Pp. 179–209. Chicago: University of Chicago Press.
Cooper, Frederick
1980 From Slaves to Squatters: Plantation Labor and Agriculture in Zanzibar and Coastal Kenya, 1890–1925. New Haven, Conn.: Yale University Press.
1989 From Free Labor to Family Allowances: Labor and African Society in Colonial Discourse. American Ethnologist 16(4): 745–756.
1992 Colonizing Time: Work Rhythms and Labor Conflict in Colonial Mombasa. *In* Colonialism and Culture. Nicholas B. Dirks, ed. Pp. 209–245. Ann Arbor: University of Michigan Press.
1996 Decolonization and African Society: The Labor Question in French and British Africa. Cambridge: Cambridge University Press.
1997 Modernizing Bureaucrats, Backward Africans, and the Development Concept. *In* International Development and the Social Sciences: Essays on the History and Politics of Knowledge. Frederick Cooper and Randall Packard, eds. Pp. 64–92. Berkeley: University of California Press.
Cooper, Frederick, and Randall Packard, eds.
1997 International Development and the Social Sciences: Essays on the History and Politics of Knowledge. Berkeley: University of California Press.
Coronil, Fernando
1996 Beyond Occidentalism: Toward Nonimperial Geohistorical Categories. Cultural Anthropology 11(1): 51–87.
1997 The Magical State: Nature, Money, and Modernity in Venezuela. Chicago: University of Chicago Press.
2000 Towards a Critique of Globalcentrism: Speculations on Capitalism's Nature. Public Culture 12(2): 351–374.
Courchamp, Franck, Elena Angulo, Philippe Rivalan, Richard J. Hall, Laetitia Signoret, Leigh Bull, and Yves Meinard
2006 Rarity Value and Species Extinction: The Anthropogenic Allee Effect. PLoS Biology 4(12): 2405–2410.
Covell, Maureen
1987 Madagascar: Politics, Economics and Society. London: Frances Pinter.
Cotte, P. V.
1946 Regardons Vivre une Tribu Malgache: Les Betsimisaraka. Paris: La Nouvelle Edition.
Crary, Jonathan
1990 Techniques of the Observer: On Vision and Modernity in the Nineteenth Century. Cambridge, Mass.: MIT Press.

Critical Ecosystem Partnership Fund
2000 Madagascar Ecosystem of the Madagascar and Indian Ocean Islands Biodiversity Hotspot. Final Version Report, December 24. http://www.cepf.net/ImageCache/cepf/content/pdfs/final_2emadagascar_2eep_2epdf/v1/final.madagascar.ep.pdf, accessed April 4, 2005.

Croll, Elizabeth, and David Parkin
1992 Bush Base: Forest Farm: Culture, Environment and Development. New York: Routledge.

Crush, Jonathan
1995 Power of Development. London: Routledge.

Daston, Lorraine, and Fernando Vidal, eds.
2003 The Moral Authority of Nature. Chicago: University of Chicago Press.

Davis, Diana K.
2007 Resurrecting the Granary of Rome: Environmental History and French Colonial Expansion in North Africa. Athens: Ohio University Press.

De Boeck, Filip
1994 Of Trees and Kings: Politics and Metaphor among the Aluund of Southwestern Zaire. American Ethnologist 21(3): 451–473.

Decary, Raymond
1958 Histoire Politique et Coloniale: Histoire des Populations Autres que les Merina, Betsileo, Betsimisaraka, Antanosy, Sihanaka, Tsimihety, Bezanozano, Antanala, Antankarana, Bara, Mahafaly, Antandroy. Vol. 3. Antananarivo, Madagascar: Imprimerie Officielle.

Dekeyser, P.-L.
1963 Principes et Historique de la Conservation de la Nature (1933–1963). Notes Africaines 99: 97–103.

Deschamps, Hubert
1960 Histoire de Madagascar. Paris: Editions Berger-Levrault.

De-Shalit, Avner
1992 Community and the Rights of Future Generations: A Reply to Robert Elliot. Journal of Applied Philosophy 9(1): 105–115.

Development Environment Consult
N.d. Filière Girofle. Rapport Module D3. Pp. 30–36.

Dewar, Robert, and David Burney
1994 Recent Research in the Paleoecology of the Highlands of Madagacar and its Implications for Prehistory. Taloha 12: 79–87.

Dewar, Robert, and Jean-Aimé Rakotoarisoa
1990 A History of Human Transformation of Madagascar. Paper delivered at the International Conference of Systematic and Evolutionary Biologists, Beltsville, Md., July.

Dorosh, Paul, and René Bernier
1994 Staggered Reforms and Limited Success: Structural Adjustment in Madagascar. *In* Adjusting to Policy Failure in African Economies. David E. Sahn, ed. Ithaca, N.Y.: Cornell University Press.

Dorr, Laurence J.
1997 Plant Collectors in Madagascar and the Comoro Islands. Washington, D.C.: Smithsonian Institution.

Dorr, Laurence J., Lisa C. Barnett, and Armand Rakotozafy
1989 Madagascar. *In* Floristic Inventory of Tropical Countries. David G. Campbell and H. David Hammond, eds. Pp. 236–50. Bronx: New York Botanical Garden.
Dove, Michael R.
1992 Foresters' Beliefs about Farmers: A Priority for Social Science Research in Social Forestry. Agroforestry Systems 17(1): 13–41.
1994 The Existential Status of the Pakistani Farmer: A Study of Institutional Factors in Development. Ethnology 33(4): 331–51.
2005 Shade: Throwing Light on Politics and Ecology in Contemporary Pakistan. *In* Political Ecology across Spaces, Scales, and Social Groups. Susan Paulson and Lisa L. Gezon, eds. Pp. 217–238. New Brunswick: Rutgers University Press.
Du Bois, Kathleen E.
1997 The Illegal Trade in Endangered Species. African Security Review 6(1): 28–41.
Duffy, Rosaleen
2008 Neoliberalising Nature: Global Networks and Ecotourism Development in Madagascar. Journal of Sustainable Tourism 16(3): 327–344.
Durbin, J., K. Bernard, and M. Fenn
1993 The Role of Socioeconomic Factors in Loss of Malagasy Biodiversity. *In* The Natural History of Madagascar. Steven M. Goodman and Jonathan P. Benstead, eds. Pp. 142–146. Chicago: University of Chicago Press.
Duvernoy, Frédéric
1987 Rapport de Mission à Mananara-Nord (Madagascar), du 1 août au 31 octobre 1987. Sous la direction du Pr. Albignac. Basançon, décembre.
Edelman, Marc
2005 Bringing the Moral Economy Back In . . . to the Study of Twenty-first Century Transnational Peasant Movements. American Anthropologist 107(3): 331–345.
Egboh, Edmond O.
1979–1980 Legal Efforts to Control Nigerian Forests Interest of the Metropolitan Economy, 1897–1940. Quarterly Review of Historical Studies 19(1–2): 64–90.
Ehrlich, Paul R., and Andrew Beattie
2001 Wild Solutions: How Biodiversity Is Money in the Bank. New Haven, Conn.: Yale University Press.
Eiss, Paul K., and David Pedersen
2002 Introduction: Values of Value. Cultural Anthropology 17(3): 283–291.
Elliot, Robert
1989 The Rights of Future People. Journal of Applied Philosophy 6(2): 159–169.
Ellis, William
1859 Three Visits to Madagascar during the Years 1853–1854–1856. New York: Harper and Brothers.
Elyachar, Julia
2005 Markets of Dispossession: NGOs, Economic Development, and the State. Durham, N.C.: Duke University Press.
Emoff, Ron
2002 Recollecting from the Past: Musical Practice and Spirit Possession on the East Coast of Madagascar. Middletown, Conn.: Wesleyan University Press.

Environment News Service
2006 Madagascar Declaration: Value of Nature Key to African Development. June 26. http://www.ens-newswire.com/ens/jun2006/2006-06-26-02.asp, accessed September 5, 2006.
Erdmann, T. K.
2003 The Dilemma of Reducing Shifting Cultivation. *In* The Natural History of Madagascar. Steven M. Goodman and Jonathan P. Benstead, eds. Pp. 134–139. Chicago: University of Chicago Press.
Escobar, Arturo
1995 Encountering Development: The Making and Unmaking of the Third World. Princeton, N.J.: Princeton University Press.
1996 Construction Nature: Elements for a Post-Structuralist Political Ecology. Futures 28(4): 325–343.
1997 Cultural Politics and Biological Diversity: State, Capital, and Social Movements in the Pacific Coast of Colombia. *In* The Politics of Culture in the Shadow of Capital. Lisa Lowe and David Lloyd, eds. Pp. 201–226. Durham, N.C.: Duke University Press.
1999 After Nature: Steps to an Antiessentialist Political Ecology. Current Anthropology 40:1–30.
n.d. Places and Regions in the Age of Globality: Social Movements and Biodiversity Conservation in the Colombian Pacific. Unpublished manuscript.
Esoavelomandroso, Manassé
1978 Religion et Politique: L'evangelisation du pays Betsimisaraka à la fin du XIX siècle. Omaly sy Anio 7–8: 7–42.
1979 La Province Maritime Orientale de la Royaume de Madagascar à la Fin du XIX Siècle. Antananarivo, Madagascar: FTM.
Evans, Ruth
2002 Madagascar Biodiversity Threatened. BBC News, January 16. http://news.bbc.co.uk/2/hi/africa/1761893.stm, accessed August 26, 2010.
Evers, Sandra J. T. M.
2002 Constructing History, Culture and Inequality: The Betsileo in the Extreme Southern Highlands of Madagascar. Leiden: Brill.
Fairhead, James, and Melissa Leach
1994 Contested Forests: Modern Conservation and Historical Land Use in Guinea's Ziama Reserve. African Affairs 93: 481–512.
1996 Misreading the African Landscape: Society and Ecology in a Forest-Savanna Mosaic. African Studies Series. Cambridge: Cambridge University Press.
Fall, Babacar
1993 Le Travail Forcé en Afrique-Occidentale Française, 1900–1946. Collection Hommes et sociétés. Paris: Karthala.
Fanony, Fulgence
1975 Fasina: Dynamisme Social et Recours a la Tradition. Travaux et Documents 14. Antananarivo, Madagascar: Musée d'Art et d'Archeologie de L'universite de Madagascar.
Fanony, Fulgence, and Henry Wright
2003 Betsimisaraka Spears from the Mananara Valley. Michigan Discussions in Anthropology 14: 53–62.

FAO [Food and Agriculture Organization of the United Nations]
2001 Global Forest Resources Assessment, 2000. Annex 3, Global Tables; Table 2. http://www.fao.org/docrep/004/Y1997E/Y1997E00.HTM, accessed December 28, 2011.
Feeley-Harnik, Gillian
1984 The Political Economy of Death: Communication and Change in Malagasy Colonial History. American Ethnologist 11(1): 1–19.
1986 Ritual and Work in Madagascar. *In* Madagascar: Society and History. Conrad P. Kottak, Jean-Aimé Rakotoarisoa, Aidan Southall, and Pierre Vérin, eds. Pp. 157–174. Durham, N.C.: Carolina Academic Press.
1991 A Green Estate: Restoring Independence in Madagascar. Washington, D.C.: Smithsonian Institution Press.
1995 Plants and People, Children or Wealth: Shifting Grounds of "Choice" in Madagascar. PoLAR 18(2): 45–64.
1999 "Communities of Blood": The Natural History of Kinship in Nineteenth Century America. Comparative Studies in Society and History 41(2): 215–262.
2001 Ravenala Madagascariensis Sonnerat: The Historical Ecology of "Flagship Species" in Madagascar. Ethnohistory 48(1–2): 31–86.
2004 The Geography of Descent. Radcliffe-Brown Lecture in Social Anthropology. Proceedings from the British Academy 125: 311–364.
Fegen, Brian
1986 Tenants' Non-violent Resistance to Landowner Claims in a Central Luzon Village. Journal of Peasant Studies 13(2): 87–106.
Felter, Harvey Wickes, and John Uri Lloyd
2004 [1898] King's American Dispensatory. Cincinnatti: Ohio Valley Co. Scanned version by Henriette Kress, 1999–2004.
Ferguson, James
1990 The Anti-politics Machine: Development, Depoliticization, and Bureaucratic Power in Lesotho. Cambridge: Cambridge University Press.
1999 Expectations of Modernity: Myths and Meanings of Urban Life on the Zambian Copperbelt. Berkeley: University of California Press.
Ferraro, Paul J.
2001 Global Habitat Protection: Limitations of Development Interventions and a Role for Conservation Performance Payments. Conservation Biology 15(4): 990–1000.
Ferraro, Paul J., and R. David Simpson
2003 Protecting Forests and Biodiversity: Are Investments in Eco-Friendly Production Activities the Best Way to Protect Endangered Ecosystems and Enhance Rural Livelihoods? Paper presented at the International Conference on Rural Livelihoods, Forests and Biodiversity, May 19–23, 2003, Bonn, Germany.
Ferry, Elizabeth Emma
2005 Not Ours Alone: Patrimony, Value, and Collectivity in Contemporary Mexico. New York: Columbia University Press.
Ferry, Elizabeth Emma, and Mandana E. Limbert
2008 Timely Assets: The Politics of Resources and Their Temporalities. Sante Fe, N.M.: School for Advanced Research Press.
Fisher, William F.
1997 Doing Good? The Politics and Antipolitics of NGO Practices. Annual Review of Anthropology 26: 439–464.

Flacourt, Etienne de
1995 [1661] Histoire de la Grande Isle de Madagascar. Edition Annotée et Presentée par Claude Allibert. Paris: Karthala.
Forest Conservation Portal
2003 Madagascar to Triple Protected Areas. September 16. http://forests.org/articles, accessed March 23, 2005.
Forsyth, T.
2003 Critical Political Ecology: The Politics of Environmental Science. London: Routledge.
Fortes, Meyer
1962 Introduction. *In* The Developmental Cycle in Domestic Groups. Jack Goody, ed. Pp. 1–14. Cambridge: Cambridge University Press.
Fortmann, Louise
1985 The Tree Tenure Factor in Agroforestry with Particular Reference to Africa. Agroforestry Systems 2: 229–251.
Foster, John Bellamy
2000 Marx's Ecology: Materialism and Nature. New York: Monthly Review Press.
2002 Ecology against Capitalism: The Nature of the Contradiction. Monthly Review Press 54(4). http://monthlyreview.org/2002/09/01/capitalism-and-ecology, accessed August 10, 2010.
Foster, John Bellamy, and Brett Clark
2004 Ecological Imperialism: The Curse of Capitalism. Socialist Register 40: 186–201.
Foster, Robert J.
2008 Commodities, Brands, Love and Kula: Comparative Notes on Value Creation in Honor of Nancy Munn. Anthropological Theory 8(9): 9–26.
Fremigacci, Jean
2007 La Vérité sur la Grande Révolte de Madagascar. L'Histoire 318: 36–43.
Freudenberger, Karen
2010 Paradise Lost? Lessons from 25 Years of USAID Environment Programs in Madagascar. Washington, D.C.: International Resources Group. http://usaid.gov/locations/sub-saharan_africa/countries/madagascar/paradise_lost_25years_env_programs.pdf, accessed December 16, 2011.
Friedland, Julian
2006 Wittgenstein and the Metaphysics of Ethical Value. Florianópolis 5(1): 91–102.
Frow, John
1997 Time and Commodity Culture: Essays on Cultural Theory and Postmodernity. Oxford: Oxford University Press.
Gade, Daniel W.
1984 Redolence and Land Use on Nosy Be, Madagascar. Journal of Cultural Geography 4(2): 29–40.
Gallieni, Joseph-Simon
1908 Neuf Ans à Madagascar. Paris: Librarie Hachette et Companie.
Garland, Elizabeth
2006 State of Nature: Colonial Power, Neoliberal Capital, and Wildlife Management in Tanzania. Ph.D. dissertation, University of Chicago.

2008 The Elephant in the Room: Confronting the Colonial Character of Wildlife Conservation in Africa. African Studies Review 51(3): 51–74.
Garner, Andrew
2004 Living History: Trees and Metaphors of Identity in an English Forest. Journal of Material Culture 9(1): 87–100.
Gerety, Rowan Moore
2009a Major International Banks, Shipping Companies, and Consumers Play Key Role in Madagascar's Logging Crisis. December 16. http://news.mongabay.com/2009/1215-rowan_madagascar.html, accessed May 5, 2010.
2009b Mining and Biodiversity Offsets in Madagascar: Conservation or "Conservation Opportunities?" August 30. http://news.mongabay.com/2009/0830-rowan_rio_tinto_madagascar.html, accessed December 14,2011.
Gezon, Lisa
1997a Institutional Structure and the Effectiveness of Integrated Conservation and Development Projects: Case Study from Madagascar. Human Organization 56(4): 462–470.
1997b Political Ecology and Conflict in Ankarana, Madagascar. Ethnology 36: 85–101.
1999 Of Shrimps and Spirit Possession: Toward a Political Ecology of Resource Management in Northern Madagascar. American Anthropologist 101(1): 58–67.
2000 The Changing Face of NGOs: Structure and Communitas in Conservation and Development in Madagascar. Urban Anthropology and Studies of Cultural Systems and World Economic Development 29(2): 181–215.
Ghimire, Krishna B.
1994 Parks and People: Livelihood Issues in National Parks Management in Thailand and Madagascar. Development and Change 25: 195–229.
Gibson, C. C, and S. A. Marks
1995 Transforming Rural Hunters into Conservationists: An Assessment of Community-Based Wildlife Management Programs in Africa. World Development 23: 941–957.
Giles-Vernick, Tamara
2002 Cutting the Vines of the Past: Environmental Histories of the Central African Rain Forest. Charlottesville: University of Virginia Press.
Ginn, Franklin
2008 Extension, Subversion, Containment: Eco-nationalism and (Post)colonial Nature in Aotearoa New Zealand. Transactions of the Institute of British Geographers 33: 335–353.
Glick, Peter
1999 Patterns of Employment and Earnings in Madagascar. Ilo Project Working Papers: Improved Policy Analysis for Economic Decision-Making and Improved Public Information Dialogue, January. Pp. 1–51. http://www.ilo.cornell.edu/ilo/workpap.html, accessed February 19, 2007.
Glick, Peter, and François Rouband
2006 Export Processing Zone Expansion in Madagascar: What Are the Labour Market and Gender Impacts? Journal of African Economies 2006 15(4): 722–756.
Global Witness and the Environmental Investigation Agency, Inc. (U.S.)
2009 Investigation into the Illegal Felling, Transport and Export of Precious Wood in Sava Region Madagascar. Conducted in cooperation with Madagascar

National Parks, the National Environment and Forest Observatory, and the Forest Administration of Madagascar, August. http://www.parcs-madagascar.com/doc/report_vsfinal.pdf, accessed December 28, 2011.

Gold, Ann Grodzins

2003 Foreign Trees: Lives and Landscapes in Rajasthan. *In* Nature in the Global South: Environmental Projects in South and Southeast Asia. Paul Greenough and Anna Lowenhaupt Tsing, eds. Pp. 170–196. Durham, N.C.: Duke University Press.

Gold, Ann Grodzins, and Bhoju Ram Guhar

2002 In the Time of Trees and Sorrows: Nature, Power, and Memory in Rajasthan. Durham, N.C.: Duke University Press.

Goldman, Michael

2004 Eco-Governmentality and Other Transnational Practices of a "Green" World Bank. *In* Liberation Ecologies: Environment, Development, Social Movements. 2nd edition. Richard Peet and Michael Watts, eds. Pp. 153–177. London: Routledge.

2005 Imperial Nature: The World Bank and Struggles for Social Justice in the Age of Globalization. New Haven, Conn.: Yale University Press.

Goldman, Michael, and Rachel A. Shurman

2000 Closing the Great Divide: New Social Theory on Society and Nature. Annual Review of Sociology 26: 563–584.

Goodman, Steven M. and Jonathan P. Benstead, eds.

2003 The Natural History of Madagascar. Chicago: University of Chicago Press.

Gow, Bonar A.

1997 Admiral Didier Ratsiraka's Revolution. Journal of Modern African Studies 35(3): 409–439.

Graeber, David

2001 Toward an Anthropological Theory of Value: The False Coin of Our Own Dreams. New York: Palgrave.

2007 Lost People: Magic and the Legacy of Slavery in Madagascar. Bloomington: Indiana University Press.

2009 Dancing with Corpses Reconsidered: An Interpretation of Famadihana (Arivonimamo, Madagascar). American Ethnologist 22(2): 258–278.

Grandidier, Alfred

1892 Histoire Physique, Naturelle et Politique de Madagascar. Paris: Imprimerie Nationale.

Grandidier, Guillaume

1920 Madagascar. Geographical Review 10(4): 197–222.

Green, Glen M., and Robert W. Sussman

1990 Deforestation History of the Eastern Rain Forests of Madagascar from Satellite Images. Science 248: 212–215.

Greenough, Paul, and Anna Lowenhaupt Tsing, eds.

2003 Nature in the Global South: Environmental Projects in South and Southeast Asia. Durham, N.C.: Duke University Press.

Grove, Richard H.

1995 Green Imperialism: Colonial Expansion, Tropical Island Edens and the Origins of Environmentalism, 1600–1860. Cambridge: Cambridge University Press.

Gruss, Capitaine
1902 Les Automobiles à Madagascar. Revue de Madagascar 1: 193–218.
GTOS [Office of the Global Terrestrial Observing System] and FAO [Food and Agriculture Organization of the United Nations]
2001 Report, Special Meeting on Biosphere Reserve Integrated Monitoring (BRIM), Rome, Italy, September 4–6.
Guichon, A.
1969 La Législation et la Réglementation de l'Exploitation Forestière à Madagascar et leur Application Pratique. Terre Malgache, Tany Malagasy 6: 137–169.
Haenn, Nora
2005 Fields of Power, Forests of Discontent: Culture, Conservation, and the State in Mexico. Tucson: University of Arizona Press.
Hall, Richard J., E. J. Milner-Gulland, and F. Courchamp
2008 Endangering the Endangered: The Effects of Perceived Rarity on Species Exploitation. Conservation Letters 1: 75–78.
Hannah, Lee, Berthe Rakotosamimanana, Jorg Ganzhorn, Rullell A. Mittermeier, Silvio Olivieri, Lata Iyer, Serve Rajaobelina, John Hough, Fanja Andriamialisoa, Ian Bowles, and Georges Tilkin
1998 Participatory Planning, Scientific Priorities, and Landscape Conservation in Madagascar. Environmental Conservation 25(1): 30–36.
Hannebique, Jacques (texte et photographies), avec la collaboration de Robert Boudry, Flavien Ranaivo, Jean-Aimé Rakotoarisoa, et Eugène Toulet
1987 Madagascar: Mon-île-au-bout-du-monde. Laval, France: Éditions Siloë.
Hanotaux, Gabriel, and Alfred Martineau
1933 Histoire des Colonies Françaises de L'Expansion de la France Dans la Monde. Paris: Société de l'Histoire Nationale.
Hanski, Ilkka, Helena Koivulehto, Alison Cameron, and Pierre Rahagalala
2007 Deforestation and Apparent Extinctions of Endemic Forest Beetles in Madagascar. Biology Letters 3(3): 344–347.
Hanson, Paul W.
1997 The Politics of Need Interpretation in Madagascar's Ranomafana National Park. Ph.D. dissertation, University of Pennsylvania, Philadelphia.
2007 Governmentality, Language Ideology, and the Production of Needs in Malagasy Conservation and Development. Cultural Anthropology 22(2): 244–284.
Harper, Janice
2002 Endangered Species: Health, Illness and Death among Madagascar's People of the Forest. Durham, N.C.: Carolina Academic Press.
Harvey, David
1990 The Condition of Postmodernity: An Enquiry into the Origins of Cultural Change. Cambridge, Mass.: Blackwell.
1996 Justice, Nature and the Geography of Difference. Cambridge, Mass.: Blackwell.
Hayden, Cori
2003 When Nature Goes Public: The Making and Unmaking of Bioprospecting in Mexico. Princeton, N.J.: Princeton University Press.
Hecht, Susanna, and Alexander Cockburn
1990 The Fate of the Forest: Developers, Destroyers, and Defenders of the Amazon. New York: HarperCollins.

Hinton, Wayne
2008 With Picks, Shovels, and Hope: The CCC and its Legacy on the Colorado Plateau. Missoula, Mont.: Mountain Press Publications.
Hobart, Charles H.
1982 Industrial Employment of Rural Indigenes: The Case of Canada. Human Organization 41(1) 54–63.
Hodgson, Dorothy L.
2001 Once Intrepid Warriors: Gender, Ethnicity, and the Cultural Politics of Maasai Development. Bloomington: Indiana University Press.
Holmes, Douglas R.
1989 Cultural Disenchantments: Worker Peasantries in Northeast Italy. Princeton, N.J.: Princeton University Press.
Hufty, Marc, and Frank Muttenzer
2002 Devoted Friends: The Implementation of the Convention on Biological Diversity in Madagascar. *In* Governing Global Biodiversity: The Evolution and Implementation of the Convention on Biological Diversity. Philippe G. Le Prestre, ed. Pp. 279–309. Burlington, Vt.: Ashgate.
Hughes, David McDermott
2005 Third Nature: Making Space and Time in the Great Limpopo Conservation Area. Cultural Anthropology 20(2): 157–184.
2006 From Enslavement to Environmentalism: Politics on a Southern African Frontier. Seattle: University of Washington Press.
Humbert, Henri
1927 Destruction d'une Flore Insulaire par le Feu: Principaux Aspects de la Végétation à Madagascar. Antananarivo, Madagascar: Imprimerie Moderne de l'Emyrne, G. Pitot et Cie.
1949 La Dégradation des Sols à Madagascar. Mémoires de l'Institut de Recherche Scientifique de Madagascar D1(1): 33–52.
Humphrey, Matthew
2002 Perservation Versus the People? Nature, Humanity, and Political Philosophy. Oxford: Oxford University Press.
Hunter, Dave
2007 Gibson Tone Tips: It All Starts with the Wood. Gibson Lifestyle. November 15. http://www.gibson.com/en-us/Lifestyle/Lessons/InstrumentLessons/Gibson%20Tone%20Tips_%20It%20All%20Start/, accessed May 5, 2010.
Huntington, Ellsworth
1924 Civilization and Climate. New Haven, Conn.: Yale University Press.
Igoe, Jim
2003 Conservation and Globalization: A Study of National Parks and Indigenous Communities from East Africa to South Dakota. Belmont, Calif.: Wadsworth.
Igoe, Jim, and Tim Kelsall
2005 Between a Rock and a Hard Place: African NGOs, Donors, and the State. Durham, N.C.: Carolina Academic Press.
Igoe, Jim, Katja Neves, and Dan Brockington
2010 A Spectacular Eco-Tour around the Historic Bloc: Theorising the Convergence of Biodiversity Conservation and Capitalist Expansion. Antipode 42(3): 486–512.

Ingold, Tim
2000 The Perception of the Environment: Essays in Livelihood, Dwelling, and Skill. London: Routledge.
2004 Culture on the Ground: The World Perceived Through the Feet. Journal of Material Culture 9(3): 315–340.
International Labour Conference
1939 Summary of Annual Reports under Article 22 of the Constitution of the ILO; Twenty-fifth Session. Geneva: International Labour Office.
International Monetary Fund
2007 Republic of Madagascar: First Review under the Three-Year Arrangement under the Poverty Reduction and Growth Facility and Request for Waiver and Modification of Performance Criteria. IMF Country Report No. 07/7. Washington, D.C.: Publication Services.
IUCN
1972 Comptes Rendus de la Conférence internationale sur la Conservation de la Nature et de ses Ressources à Madagascar, Tananarive 7–11 Octobre, 1970. Publications UICN Nouvelle Série, Document Supplémentaire 36. Morges, Switzerland: IUCN.
Jacoby, Karl
2001 Crimes against Nature: Squatters, Poachers, Thieves, and the Hidden History of American Conservation. Berkeley: University of California Press.
Jarosz, Lucy
1993 Defining and Explaining Tropical Deforestation: Shifting Cultivation and Population Growth in Colonial Madagascar (1896–1940). Economic Geography 69(4): 366–379.
Jennings, Eric T.
2003 Remembering "Other" Losses: The Temple du Souvenir Indochinois of Nogent-sur-Marne. History and Memory 15(1): 5–48.
2006 Curing the Colonizers: Hydrotherapy, Climatology, and French Colonial Spas. Durham, N.C.: Duke University Press.
Jones, Julia P. G., Mijasoa M. Andriamarovololona, and Neal Hockley
2008 The Importance of Taboos and Social Norms to Conservation in Madagascar. Conservation Biology 22(4): 976–986.
Journal Officiel de Madagascar et Dépendances
1939 Arrêté Dans la Colonie de Madagascar et Dépendances, le Décret du 31 Mai 1938 portant ratification de la convention internationale pour la protection de la faune et de la flore en Afrique Signée à Londres le 8 Novembre 1933. April 3. No. 644.
Kaufman, Herbert
2006 [1960] The Forest Ranger: A Study in Administrative Behavior. Washington, D.C.: Resources for the Future.
Kaufmann, Jeffrey C.
2001 La Question des Raketa: Colonial Struggles with Prickly Pear Cactus in Southern Madagascar, 1900–1923. Ethnohistory 48(1–2): 87–122.
Kaufmann, Jeffrey C., ed.
2008 Greening the Great Red Island: Madagascar in Nature and Culture. Pretoria: Africa Institute of South Africa.

Keller, Eva

2008 The Banana Plant and the Moon: Conservation and the Malagasy Ethos of Life in Masoala, Madagascar. American Ethnologist 35(4): 650–664.

2009 The Danger of Misunderstanding "Culture." Madagascar Conservation and Development 4(2): 82–85.

King, Robert Thomas

1991 The Free Life of a Ranger: Archie Murchie in the U.S. Forest Service, 1929–1965. Reno: University of Nevada Oral History Program.

Kirsch, Stuart

2006 Reverse Anthropology: Indigenous Analysis of Social and Environmental Relations in New Guinea. Stanford, Calif.: Stanford University Press.

Klein, Jorgen

2002 Deforestation in the Madagascar Highlands: Established "Truth" and Scientific Uncertainty. GeoJournal 56: 191–199.

Koehler, Robert

2009 It's Not Imperialism! It's a Uniquely Korean Dream! The Marmot's Hole, February 18. http://www.rjkoehler.com/2009/02/18/its-not-imperialism-its-a-uniquely-korean-dream/, accessed December 28, 2011.

Kottak, Conrad

1999 The New Ecological Anthropology. American Anthropologist 101(1): 23–35.

Kramer, Randall A., Narendra Sharma, and Mohan Munasinghe

1995 Valuing Tropical Forests: Methodology and Case Study of Madagascar. Environmental Paper. Washington, D.C.: World Bank.

Kremen, Claire, Adina M. Merenlender, and Dennis D. Murphy

1994 Ecological Monitoring: A Vital Need for Integrated Conservation and Development Programs in the Tropics. Conservation Biology 8(2): 388–397.

Kuhlken, Robert

1999 Settin' the Woods on Fire: Rural Incendiarism as Protest. Geographical Review 89(3): 343–363.

Kull, Christian

1996 The Evolution of Conservation Efforts in Madagascar. International Environmental Affairs 8(1): 50–86.

2002 Madagascar Aflame: Landscape Burning as Peasant Protest, Resistance, or a Resource Management Tool? Political Geography 21(7): 927–953.

2004 Isle of Fire: The Political Ecology of Landscape Burning in Madagascar. Chicago: University of Chicago Press.

Lahady, Pascal

1979 Le Culte Betsimisaraka et Son Système Symbolique. Fianarantsoa, Madagascar: Librairie Ambozontany.

Lakoff, George, and Mark Johnson

1980 Metaphors We Live By. Chicago: University of Chicago Press.

Lambek, Michael

2000 Nuriaty, the Saint and the Sultan: Virtuous Subject and Subjective Virtuoso of the Post-modern Colony. Anthropology Today 16(2): 8–12.

Landau, Peter

1999 Prices for Cloves Skyrocket; No Relief until the New Crop. Chemical Market Reporter 256(24): July 19.

Larson, B. A.
1994 Changing the Economics of Environmental Degradation in Madagascar: Lessons from the National Environmental Action Plan process. World Development 22: 671–689.
Larson, Pier
1996 Desperately Seeking "the Merina" (central Madagascar): Reading Ethnonyms and Their Semantic Fields in African Identity Histories. Journal of Southern African Studies 13(2): 541–560.
Latour, Bruno
1999 Pandora's Hope: An Essay on the Reality of Science Studies. Cambridge, Mass.: Harvard University Press.
Lavauden, L.
1931 Le Problème Forestière a Madagascar. Communication Presentée au Congrès de la Production Forestière Coloniale et Nord-Africaine Tenu à l'Occasion de l'Exposition Coloniale Internationale, Paris. Paris: Exposition Coloniale Internationale de Paris.
Lawrence, Geoffrey, Higgins Vaughan, and Stewart Lockie, eds.
2001 Environment, Society and Natural Resource Management: Theoretical Perspectives from Australasia and the Americas. Cheltenham, UK: Edward Elgar.
Leach, James
2003 Creative Land: Place and Procreation on the Rai Coast of Papua New Guinea. New York: Berghahn Books.
Leach, Melissa, and Robin Mearns, eds.
1996 The Lie of the Land: Challenging Received Wisdom on the African Environment. London: Heinemann.
Leakey, Richard, and Roger Lewin
1996 The Sixth Extinction: Biodiversity and Its Survival. London: Weidenfeld and Nicolson.
Leblond, Marius-Ary
1907 La Grande Ile de Madagascar. Paris: Librarie CH. Delagrave.
Lefebvre, Henri
1991 [1974] The Production of Space. Oxford: Blackwell.
Leff, Enrique
1995 Green Production: Toward an Environmental Rationality. New York: Guildford Press.
Lele, Uma, and Christopher Gerrard
2003 Global Public Goods, Global Programs, and Global Policies: Some Initial Findings from a World Bank Evaluation. American Journal of Agricultural Economics 85(3): 686–691.
Leopold, Aldo
1989 [1949] A Sand County Almanac Illustrated. New York: Oxford University Press.
Leymaire, Philippe
1975 Le Fokonolona: La Voie Malgache du Socialisme. Revue Française d'Etudes Politiques Africaines 112: 47–53.
Lien, Marianne E.
2005 "King of Fish" or "Feral Peril": Tasmanian Atlantic Salmon and the Politics of Belonging. Environment and Planning D: Society and Space 23: 659–671.

Light, Andrew, and Holmes Rolston, eds.
2002 Environmental Ethics. Boston, Mass.: Blackwell.
Lind, J. R.
2009 Feds Raid Gibson Offices. Nashville Post. November 17. http://nashvillepost.com/news/2009/11/17/feds_raid_gibson_offices, accessed December 1, 2009.
Littlefield, Alice
1978 Exploitation and the Expansion of Capitalism: The Case of the Hammock Industry of Yucatan. American Ethnologist 5(3): 495–508.
Lloyd, J. A.
1850 Memoir on Madagascar. Journal of the Royal Geographical Society of London 20: 53–75.
Locatelli, Bruno
2000 Pression Démographique et Construction du Paysage Rural des Tropiques Humides l'Exemple de Mananara (Madagascar). Ph.D. dissertation, Ecole National du Génie Rural, des Eaux et Fôrets, Centre de Monpellier.
Lohmann, Larry
1993 Green Orientalism. Economist 23(6): 202–204.
Lomnitz, Claudio
1994 Decadence in Times of Globalization. Cultural Anthropology 9(2): 257–267.
Lonifavski, M.
1922 L'Avenir de la Colonisation à Madagascar. Le Domaine—Les Concessions. Bulletin Economique (1er Trimestre): 61–62.
Louvel, M.
1929 Lutte Contre le Feu de Brousse par le Reboisement. Bulletin de L'Academie Malgache. No. 1 Partie: Documentation Etudée Madagascar et Dependances.
1952 Les Reboisements. Bulletin de l'Académie Malgache, Numéro Special du Cinquantenaire: 43–51.
1954 La Vieille Forêt Malgache. Bulletin de l'Académie Malgache 32: 76–78.
Lowe, Celia
2006 Wild Profusion: Biodiversity Conservation in an Indonesian Archipelago. Princeton, N.J.: Princeton University Press.
MacKenzie, John M.
1988 The Empire of Nature: Hunting, Conservation and British Imperialism. Manchester: Manchester University Press.
MacKenzie, John M., ed.
1990 Imperialism and the Natural World. Manchester, UK: Manchester University Press.
Madagascar Agricultural Production. www.mongabay.com, accessed August 26, 2010.
Malaivandy, Longo
2000 "Republique Humaniste et Ecologique" Hoe? Na Eritreritra Nateraky Ny Kabarin'Ny Filoha, Ny Amiraly Didier Ratsiraka Tamin'ny 9 February 1997. Antananarivo, Madagascar: DLE.
Malinowski, Bronislaw
1965 [1935] Soil-Tilling and Agricultural Rites in the Trobriand Islands: Coral Gardens and Their Magic, vol. 1. Bloomington: Indiana University Press.
Maminirina, C. P., P. Girod, and P. O. Waeber
2006 Comic Strips as Environmental Educative Tools for the Alaotra Region. Madagascar Conservation and Development 1: 11–14.

Mangalaza, Eugène Régis
1976–1979 Essai de Philosophie Betsimisaraka Sens du Famadihana. Tuléar, Madagascar: Centre Universitaire Régional de Tuléar.
1985 Les vivants à l'écoute de leurs morts: exemple des "Betsimisaraka-Antavaratra." Cahiers Ethnologiques 6: 33–50.
1994 La Poule de Dieu: Essai d'Anthropologie Philosophique Chez les Betsimisaraka. Mémoires des Cahiers Ethnologiques 4. Bordeaux: Les Presses Universitaires de Bordeaux.
Marcus, Richard R.
2001 Seeing the Forest for the Trees: Integrated Conservation and Development Projects and Local Perceptions of Conservation in Madagascar. Human Ecology 29(4): 381–397.
Marcus, Richard R., and Christian Kull
1999 Setting the Stage: The Politics of Madagascar's Environmental Efforts. African Studies Quarterly 3(2). http://web.africa.ufl.edu/asq/v3/v3i2.htm, accessed March 4, 2007.
Marx, Karl
1964 [1844] Economic and Philosophic Manuscripts of 1844. Dirk J. Struik, ed., and Martin Milligan, trans. New York: International Publishers.
1970 [1845/46] The German Ideology, Part One with Selections from Parts Two and Three and Supplementary Texts. C. J. Arthur, ed. New York: International Publishers.
1977 [1867] Capital: A Critique of Political Economy, vol. 1. Introduced by Ernest Mandel. Ben Fowkes, trans. New York: Vintage Books.
Masquelier, Adeline
2002 Road Mythographies: Space, Mobility, and the Historical Imagination in Postcolonial Niger. American Ethnologist 29(4): 829–856.
Maude, Francis Cornwallis
2011 [1895] Five Years in Madagascar, with Notes on the Military Situation. London: British Library, Historical Print Editions.
Mauss, Marcel
1990 [1950] The Gift: The Form and Reason for Exchange in Archaic Societies. W. D. Halls, trans. New York: W. W. Norton.
Mazzaro, Danielle
2000 Clove Leaf Oil Market Projected to Remain Stable. Chemical Market Reporter 258(9), August 28.
McAfee, K.
1999 Selling Nature to Save It? Society and Space 17: 133–154.
McConnell, William J., and Sean P. Sweeney
2005 Challenges of Forest Governance in Madagascar. Geographical Journal 171(2): 223–238.
McCoy, L. and H. Razafinrainibe
1997 Madagascar's Integrated Conservation and Development Projects: Lessons Learned by Participants, Project Employees, Related Authorities, and Community Beneficiaries. Antananarivo, Madagascar: USAID.
Metz, Helen Chapin, ed.
1994 Madagascar: A Country Study. Washington: GPO for the Library of Congress.

Mezger, Max
1936 [1931] Monica Goes to Madagascar. Maida C. Darnton, trans. New York: Coward-McCann.
Middleton, Karen
1999a Ancestors, Power, and History in Madagascar. Leiden: Brill.
1999b Who Killed "Malagasy Cactus"? Science, Environment and Colonialism in Southern Madagascar (1924–1930). Journal of Southern African Studies 25(2): 215–248.
Miller, Daniel
2005 Materiality: An Introduction. *In* Materiality. Daniel Miller, ed. Pp. 1–50. Durham, N.C.: Duke University Press.
Milton, Kay
2002 Loving Nature: Towards an Ecology of Emotion. New York: Routledge.
Minge-Kalman, Wanda
1978 Household Economy during the Peasant-to-Worker Transition in the Swiss Alps. Ethnology 17: 183–196.
Minten, Bart
2003 Production, Revenue Agricole et Pauvreté. *In* Agriculture, Pauvreté Rurale et Politiques Economiques à Madagascar. Bart Minten, Jean-Claude Randrianarisoa, and Lalaina Randrianarison, eds. Pp. 52–55. Part of the Ilo Project Working Papers, Improved Policy Analysis for Economic Decision-Making and Improved Public Information Dialogue. Ithaca, N.Y.: USAID, Cornell University, INSTAT, FOFIFA.
Minten, Bart, and Lalaina Randrianarison
2003 La Main-d'Oeuvre Agricole. *In* Agriculture, Pauvreté Rurale et Politiques Economiques à Madagascar. Bart Minten, Jean-Claude Randrianarisoa, and Lalaina Randrianarison, eds. Pp. 1–10. Part of the Ilo Project Working Papers, Improved Policy Analysis for Economic Decision-Making and Improved Public Information Dialogue. USAID, Cornell University, INSTAT, FOFIFA.
Mintz, Sidney W.
1985 Sweetness and Power: The Place of Sugar in Modern History. New York: Penguin Books.
Moore, Donald S.
1993 Contesting Terrains in Zimbabwe's Eastern Highlands: Political Ecology, Ethnography, and Peasant Resource Struggles. Economic Geography 69: 390–401.
1998 Subaltern Struggles and the Politics of Place: Remapping Resistance in Zimbabwe's Eastern Highlands. Cultural Anthropology 13(3): 344–381.
Moore, Donald S., Jake Kosek, and Anand Pandian
2003 The Cultural Politics of Race and Nature: Terrains of Power and Practice. *In* Race, Nature, and the Politics of Difference. Donald S. Moore, Jake Kosek, and Anand Pandian, eds. Pp. 1–70. Durham, N.C.: Duke University Press.
Moore, Henrietta L., and Megan Vaughan
1994 Cutting Down Trees: Women, Nutrition and Agricultural Change in the Northern Province of Zambia, 1890–1990. Portsmouth, N.H.: Heinemann.
Mudimbe, V. Y.
1994 The Idea of Africa. Bloomington: Indiana University Press.

Mueggler, Erik
2005 The Lapponicum Sea: Matter, Sense, and Affect in the Botanical Exploration of Southwest China and Tibet. Comparative Studies in Society and History 47(3): 442–479.
Mukonoweshuro, Eliphas G.
1990 State "Resilience" and Chronic Political Instability in Madagascar. Canadian Journal of African Studies 24(3): 376–398.
Mullens, Joseph
1876 On the Origin and Progress of the People of Madagascar. Journal of the Anthropological Institute of Great Britain and Ireland 5: 181–198.
Munn, Nancy
1983 Gawan Kula: Spatiotemporal Control and the Symbolism of Influence. *In* The Kula: New Perspectives on Massim Exchange. J. Leach and E. Leach, eds. Cambridge: Cambridge University Press.
1986 The Fame of Gawa: Value Transformation in a Massim (Papua New Guinea) Society. Cambridge: Cambridge University Press.
Munro, William A.
1998 The Moral Economy of the State: Conservation, Community Development, and State Making in Zimbabwe. Athens: Ohio University Center for International Studies.
Mutibwa, P. M.
1974 The Malagasy and the Europeans: Madagascar's Foreign Relations, 1861–1895. Atlantic Highlands, N.J.: Humanities Press.
Myers, Norman
1988 Threatened Biotas: "Hot Spots" in Tropical Forests. Environmentalist 8(3): 187–208.
Myers, Norman, Russell A. Mittermeier, Cristina G. Mittermeier, Gustavo A. B. da Fonseca, and Jennifer Kent
2000 Biodiversity Hotspots for Conservation Priorities. Nature 403: 853–858.
Nabhan, Gary Paul
1995 The Dangers of Reductionism in Biodiversity Conservation. Conservation Biology 9(3): 479–481.
Nederveen Pieterse, Jan
2002 Development Theory: Deconstructions/Reconstructions. Thousand Oaks, Calif.: Sage.
Nemours, Charles Phillippe, Duc de
1930 Madagascar et Ses Richesses. Paris: P. Roger.
Neumann, Roderick P.
1996 Dukes, Earls, and Ersatz Edens: Aristocratic Nature Preservationists in Colonial Africa. Environment and Planning D: Society and Space 14: 79–98.
1997 Primitive Ideas: Protected Area Buffer Zones and the Politics of Land in Africa. Development and Change 28(3): 559–582.
1998 Imposing Wilderness: Struggles over Livelihood and Nature Preservation in Africa. Berkeley: University of California Press.
2004 Moral and Discursive Geographies in the War for Biodiversity in Africa. Political Geography 23(7): 813–849.
New York Times
1883 Through Madagascar. June 14: 2.

1891 Madagascar's Forced Labor. From the London Daily News. November 22: 2.
1971 For Madagascar, A Pragmatic Line. January 29: 67.
1973 Madagascar: A Fresh Start, New Hopes: Efforts and Commitment to Development are Renewed. February 4: 176.

Nicoll, Martin E., and Olivier Langrand
1989 Madagascar: Revue de la conservation et des aires protegées. Gland, Switzerland: World Wide Fund for Nature.

Oates, John
1999 Myth and Reality in the Rain Forest: How Conservation Strategies Are Failing in West Africa. Berkeley: University of California Press.

O'Connor, James
1994 Is Sustainable Capitalism Possible? *In* Is Capitalism Sustainable? Political Economy and the Politics of Ecology. Martin O'Connor, ed. Pp. 152–175. New York: Guilford Press.
1998 Natural Causes: Essays in Ecological Marxism. New York: Guilford Press.

O'Connor, Martin
1994a On the Misadventures of Capitalist Nature. *In* Is Capitalism Sustainable? Political Economy and the Politics of Ecology. Martin O'Connor, ed. Pp. 125–151. New York: Guilford Press.

O'Connor, Martin, ed.
1994b Is Capitalism Sustainable? Political Economy and the Politics of Ecology. New York: Guilford Press.

Olivier, Marcel
1931 Six Ans de Politique Sociale à Madagascar. Paris: B. Grasset.

Ollman, Bertell
1976 Alienation: Marx's Conception of Man in Capitalist Society. Cambridge: Cambridge University Press.

Olson, Sherry H.
1984 The Robe of the Ancestors: Forests in the History of Madagascar. Journal of Forest History 28(4): 174–186.

Ong, Aihwa
1987 Spirits of Resistance and Capitalist Discipline: Factory Women in Malaysia. Albany: State University of New York Press.

Ong, Aihwa, and Stephen J. Collier, eds.
2005 Global Assemblages: Technology, Politics, and Ethics as Anthropological Problems. Malden, Mass.: Blackwell Science.

Orlove, Benjamin S., and Stephen B. Brush
1996 Anthropology and the Conservation of Biodiversity. Annual Review of Anthropology 25: 329–352.

Ortner, Sherry B.
1973 On Key Symbols. American Anthropologist 75(5): 1338–1346.
1995 Resistance and the Problem of Ethnographic Refusal. Comparative Studies in Society and History 37(1): 173–193.

Ouellette, Jennifer
2000 Fears of Clove Oil Shortage Could Have Little Impact. Chemical Market Reporter 257(23): June 5.

Owen, John
1962 The National Parks of Tanganyika. *In* First World Conference on National Parks. A. B. Adams, ed. Pp. 52–59. Washington, D.C.: U.S. Government Printing Office.
Padmore, George
1947 Madagascar Fights for Freedom. Left 132. October. Transcribed by Christian Hogsbjerg for the Marxists' Internet Archive 2007. http://www.marxists.org/archive/padmore/1947/madagascar.htm, accessed August 11, 2010.
Pálsson, Gísli P., and Paul Rabinow
2005 The Iceland Controversy: Reflections on the Transnational Market of Civic Virtue. *In* Global Assemblages: Technology, Politics, and Ethics as Anthropological Problems. Aihwa Ong and Stephen J. Collier, eds. Pp. 91–104. Malden, Mass.: Blackwell Science.
Pankhurst, Donna
1996 Similar but Different? Assessing the Reserve Economy Legacy of Namibia. Journal of Southern African Studies 22(3): 405–420.
Peet, Richard
1985 The Social Origins of Environmental Determinism. Annals of the Association of American Geographers 73(3): 309–333.
Peet, Richard, and Michael Watts, eds.
1993 Introduction: Development Theory and Environment in an Age of Market Triumphalism. Economic Geography 69(3): 227–253.
2004 [1996] Liberation Ecologies: Environment, Development, Social Movements. 2nd edition. London: Routledge.
Peluso, Nancy
1992 Rich Forests, Poor People: Resource Control and Resistance in Java. Berkeley: University of California Press.
1996 Fruit Trees and Family Trees in an Anthropogenic Forest: Ethics of Access, Property Zones, and Environmental Change in Indonesia. Comparative Studies in Society and History 38(3): 510–549.
Perrier de la Bâthie, Henri
1921 La Végétation Malgache. Annales du Musée Colonial de Marseille 9: 1–266.
Peters, Joe
1998 Transforming the Integrated Conservation and Development Project (ICDP) Approach: Observations from the Ranomafana National Park Project, Madagascar. Journal of Agricultural and Environmental Ethics 11: 17–47.
Peterson, Richard B.
2000 Conversations in the Rainforest: Culture, Values, and the Environment in Central Africa. Boulder, Colo.: Westview Press.
Pigg, Stacey Leigh
1992 Inventing Social Categories through Place: Social Representations and Development in Nepal. Comparative Studies in Society and History 34(3): 491–513.
Phillips, Anne
1989 The Enigma of Colonialism. Bloomington: Indiana University Press.
Platt, John
2008 Red List of Endangered Species: Bleak. Plenty, October. http://www.mnn

.com/earth-matters/wilderness-resources/stories/red-list-of-endangered-species-bleak, accessed August 23, 2010.

Polanyi, Karl
1957 [1944] The Great Transformation: The Political and Economic Origins of Our Time. Boston: Beacon Press.

Pollini, Jacques
2010 Environmental Degradation Narratives in Madagascar: From Colonial Hegemonies to Humanist Revisionism. Geoforum 41(5): 711–722.

Porter, C. R.
1940 Madagascar in 1939. The Geographical Journal 95(3): 191–207.

Potts, Deborah
2000 Worker-Peasants and Farmer-Housewives in Africa: The Debate about "Committed" Farmers, Access to Land and Agricultural Production. Journal of Southern African Studies 26(4): 807–832.

Pratt, Mary Louise
1992 Imperial Eyes: Travel Writing and Transculturation. New York: Routledge.

Rabearivelo, Jean-Joseph
1928 Volumes. Antananarivo, Madagascar: Imprimerie de l'Imerina.

Raffles, Hugh
2002 In Amazonia: A Natural History. Princeton, N.J.: Princeton University Press.

Rahonintsoa, Elyane Tiana
1978 Sainte-Marie de Madagascar: Insularité et Economie du Girofle. Masters thesis, Géographie, University of Madagascar, Antananarivo.

Raison-Jourde, Françoise
1991 Bible et Pouvoir à Madagascar au XIXe Siècle: Invention d'une Identité Chrétienne et Construction de l'Etat (1780–1880). Paris: Karthala.

Ralison, Eliane, and Bart Minten
2003 Accès aux Ressources Halieutiques et Place de la Pêche dans l'Economie Rurale. *In* Agriculture, Pauvreté Rurale et Politiques Economiques à Madagascar. Bart Minten, Jean-Claude Randrianarisoa, and Lalaina Randrianarison, eds. Pp. 64–65. Part of the Ilo ProjectWorking Papers, Improved Policy Analysis for Economic Decision-Making and Improved Public Information Dialogue. USAID, Cornell University, INSTAT, FOFIFA.

Randriamalala, Hery, and Zhou Lui
2010 Rosewood of Madagascar: Between Democracy and Conservation. Madagascar Conservation and Development 5(1): 11–22.

Randrianarison, Lalaina
2003 Revenus Extra-Agricoles des Ménages Ruraux et Pauvreté. *In* Agriculture, Pauvreté Rurale et Politiques Economiques à Madagascar. Bart Minten, Jean-Claude Randrianarisoa, and Lalaina Randrianarison, eds. Pp. 56–59. Part of the Ilo Project Working Papers, Improved Policy Analysis for Economic Decision-Making and Improved Public Information Dialogue. USAID, Cornell University, INSTAT, FOFIFA.

Ramanantsoavina, M. Georges
1966 La Forêt Malgache. Annales Malgaches Droit 3: 43–65.

Randrianja, Solofo
N.d. Le Parti Communiste de la Région de Madagascar, 1930–1939. Antananarivo, Madagascar: Foi et Justice, serie recherches historiques.
Raony-Le Boubennec, Fandresena
1999 Environnement, Quel Heritage Pour les Generations Malgaches du Troisième Millénaire. Madagascar Magazine 16: 14–16.
Rausser, Gordon C., and Arthur A. Small
2000 Valuing Research Leads: Bioprospecting and the Conservation of Genetic Resources. Journal of Political Economy 108(1): 173–206.
Raveloarinoro, Marie Gisèle
2006 L'Environnement et la Conservation Dans le Fonds Grandidier du Parc Botanique et Zoologique de Tsimbazaza. Taloha: Revue Scientifique Internationale des Civilization 16–17. http://www.taloha.info/document.php?id=365.
Razafy Fara, Lala
1998 La Forêt et La Population Locale ou la Gestion Locale des Forêts a Vohidrazana. *In* Les Stratégies Endogènes et La Gestion Des Ressources Naturelles dans la Région De Beforona: Résultats des Recherches Pluridisciplinaires de la Phase 1995 à 1998. Cahiers Terre-Tany 8. Antananarivo, Madagascar: Terre-Tany/BEMA, CDE/GIUB and FOFIFA.
Redfield, Peter
2000 Space in the Tropics: From Convicts to Rockets in French Guiana. Berkeley: University of California Press.
Redfield, R., R. Linton, and M. J. Herskovits
1936 Memorandum on the Study of Acculturation. American Anthropologist 38: 149–152.
Reitel, François
1994 Le Role de l'Armée dans la Conservation des Forêts. *In* Forêt et Guerre. Andrée Corvol and Jean-Paul Amat, eds. Pp. 49–55. Paris: Editions L'Harmattan.
Revill, Jo
2005 Madagascar's Unique Forest under Threat. Observer UK, August 6. http://www.guardian.co.uk/science/2005/aug/07/conservationandendangeredspecies.internationalnews, accessed December 6, 2011.
Ribot, Jesse C.
1995 From Exclusion to Participation: Turning Senegal's Forestry Policy Around? World Development 23(9): 1597–1599.
1998 Decentralization, Participation, and Accountability in Sahelian Forestry: Legal Instruments of Political-Administrative Control. Working Paper #00-5. Berkeley Workshop on Environmental Politics. Berkeley: Institute of International Studies, University of California.
Richards, Audrey
1939 Land, Labour and Diet in Northern Rhodesia: An Economic Study of the Bemba Tribe. London: Oxford University Press.
Rival, Laura, ed.
1998 The Social Life of Trees: Anthropological Perspectives on Tree Symbolism. New York: Berg.

Roberts, Stephen H.
1929 History of French Colonial Policy (1870–1925), vol. 2. London: P. S. King and Son.
Robbins, Paul
2004 Political Ecology: A Critical Introduction. Malden, Mass.: Blackwell Press.
Rocheleau, Dianne, and David Edmunds
1997 Women, Men and Trees: Gender, Power and Property in Forest and Agrarian Landscapes. World Development 25(8): 1351–1376.
Rocheleau, Dianne, and Laurie Ross
1995 Trees as Tools, Trees as Text: Struggles over Resources in Zambrana-Chacuey, Dominican Republic. Antipode 27(4): 407–429.
Rocheleau, Dianne, Barbara Thomas-Slayter, and Esther Wangari, eds.
1996 Feminist Political Ecology: Global Issues and Local Experiences. London: Routledge.
Sahlins, Peter
1994 Forest Rites: The War of the Desmoiselles in Nineteenth-Century France. Cambridge, Mass.: Harvard University Press.
Said, Edward W.
1979 Orientalism. New York: Vintage.
Salmond, Anne
1983 Theoretical Landscapes: On Cross Cultural Conceptions of Knowledge. *In* Semantic Anthropology. David Parkin, ed. Pp. 65–88. London: Academic Press.
Sarkar, Saral
1999 Eco-Socialism or Eco-Capitalism? A Critical Analysis of Humanity's Fundamental Choices. London: Zed Books.
Schmink, Marianne, and Charles H. Wood
1992 Contested Frontiers in Amazonia. New York: Columbia University Press.
Schroeder, Richard A.
1999 Shady Practices: Agroforestry and Gender Politics in the Gambia. Berkeley: University of California Press.
Scott, James C.
1976 The Moral Economy of the Peasant: Rebellion and Subsistence in Southeast Asia. New Haven, Conn.: Yale University Press.
1985 Weapons of the Weak: Everyday Forms of Peasant Resistance. New Haven, Conn.: Yale University Press.
1998 Seeing Like a State: How Certain Schemes to Improve the Human Condition Have Failed. New Haven, Conn.: Yale University Press.
2005 Afterward to "Moral Economies, State Spaces and Categorical Violence." American Anthropologist 107(3): 395–402.
Serrano, Richard
2005 Against the Postcolonial: Francophone Writers at the Ends of the French Empire. Lanham, Md.: Lexington Books.
Sharp, Lesley A.
1993 The Possessed and the Dispossessed: Spirits, Identity, and Power in a Madagascar Migrant Town. Berkeley: University of California Press.
2001 Wayward Pastoral Ghosts and Regional Xenophobia in a Northern Madagascar Town. Africa 71(1): 38–81.

2002 The Sacrificed Generation: Youth, History, and the Colonized Mind in Madagascar. Berkeley: University of California Press.
2003 Laboring for the Colony and Nation: The Historicized Political Consciousness of Youth in Madagascar. Critique of Anthropology 23(1) (2003): 75–91.
Shifferes, Steve
2005 Bush Treads His Own Path on Africa. BBC News, June 7. http://news.bbc.co.uk/1/hi/business/4070036.stm, accessed June 13, 2005.
Shoumatoff, Alex
1988 Our Far Flung Correspondents (Madagascar). Dispatches from the Vanishing World. http://www.dispatchesfromthevanishingworld.com/pastdispatches/madagascar/printermadagascar.html, accessed May 12, 2010.
Shue, Henry
1980 Basic Rights. Princeton, N.J.: Princeton University Press.
Sibree, James
1870 Madagascar and Its People. London: W. Clowes and Sons.
1884 Notes on Relics of the Sign and Gesture Language among the Malagasy. Journal of the Anthropological Institute of Great Britain and Ireland 13: 174–183.
1892 Curious Words and Customs Connected with Chieftainship and Royalty among the Malagasy. Journal of the Anthropological Institute of Great Britain, 21: 215–230.
Sivaramakrishnan, K.
1995 Colonialism and Forestry in India: Imagining the Past in Present Politics. Comparative Studies in Society and History 37(1): 3–40.
1997 A Limited Forest Conservancy in Southwest Bengal, 1864–1912. Journal of Asian Studies 56(1): 75–112.
1999 Modern Forests: Statemaking and Environmental Change in Colonial Eastern India. Stanford, Calif.: Stanford University Press.
2000 Crafting the Public Sphere in the Forests of West Bengal: Democracy, Development, and Political Action. American Ethnologist 27(2): 431–461.
2005 Some Intellectual Genealogies for the Concept of Everyday Resistance. American Anthropologist 107(3): 346–355.
Slater, Candace
1996 Amazonia as Edenic Narrative. *In* Uncommon Ground. William Cronon, ed. Pp. 114–131. New York: W. W. Norton
2002 Entangled Edens: Visions of the Amazon. Berkeley: University of California Press.
Smith, David G.
2000 Moral Geographies: Ethics in a World of Difference. Edinburgh: Edinburgh University Press.
Smith, Neil
1990 [1984] Uneven Development. Cambridge, Mass.: Basil Blackwell.
2007 Nature as Accumulation Strategy. *In* Coming to Terms with Nature: Socialist Register 2007. Leo Panitch and Colin Leys, eds. Pp. 16–37. New York: Monthly Review Press.
Sodikoff, Genese
2003 The Case of the Lace Leaf: 19th Century Naturalism and the Containment of Malagasy Species. Michigan Discussions in Anthropology 14: 167–192.

2004 Land and Languor: Ethical Imaginations of Work and Forest in Northeast Madagascar. History and Anthropology 15(4): 367–398.
2005a Forced and Forest Labor in Colonial Madagascar, 1926–1936. Ethnohistory 52(2): 407–435.
2005b Reserve Labor: A Moral Ecology of Conservation in Madagascar. Ph.D. dissertation, University of Michigan, Ann Arbor.
2007 An Exceptional Strike: A Micro-History of "People versus Park" in Madagascar. Journal of Political Ecology 14: 10–33.
2011 Totem and Taboo Reconsidered: Endangered Species and Moral Practice in Madagascar. *In* The Anthropology of Extinction: Essays on Culture and Species Death. Genese Marie Sodikoff, ed. Pp. 67–86. Bloomington: Indiana University Press.

Song, Jung-a, Christian Oliver, and Tom Burgis
2008 Daewoo to Cultivate Madagascar Land for Free. Financial Times, November 19. http://www.ft.com/intl/cms/s/0/6e894c6a-b65c-11dd-89dd-0000779fd18c.html#axzz1hwzHgDVP, accessed November 25, 2008.

Spectator
1895 Colonel Maude on Madagascar. August 24. Vol. 75: 244–245.

Spencer, Richard
2008 South Korean Company Takes Over Part of Madagascar to Grow Biofuels. Telegraph, November 20. http://www.telegraph.co.uk/earth/agriculture/3487668/South-Korean-company-takes-over-part-of-Madagascar-to-grow-biofuels.html, accessed January 5, 2009.

Sterba, James P.
2005 Global Justice for Humans or for All Living Beings and What Difference it Makes. Journal of Ethics 9: 283–300.

Sterba, James P., ed.
1984 Morality in Practice. Belmont, Calif.: Wadsworth.

Stevenson, Richard W.
2005 Africans Press Bush to Speed Aid Program. New York Times, June 14: A7.

Stoler, Ann Laura
1985 Capitalism and Confrontation in Sumatra's Plantation Belt, 1870–1979. New Haven, Conn.: Yale University Press.
1986 Plantation Politics and Protest on Sumatra's East Coast. Journal of Peasant Studies 13(2): 124–143.
2008 Imperial Debris: Reflections on Ruin and Ruination. Cultural Anthropology 23(2): 191–219.

Strathern, Marilyn
1992 After Nature: English Kinship in the Late Twentieth Century. Cambridge: Cambridge University Press.

Stratton, Arthur
1964 The Great Red Island. New York: Charles Scribner's Sons.

Sullivan, Sian
2010 The Environmentality of "Earth Incorporated": On Contemporary Primitive Accumulation and the Financialisation of Environmental Conservation. Paper presented at the conference An Environmental History of Neoliberalism, Lund University, Lund, Sweden, May 6–8.

Sussman, Robert W., Glen M. Green, and Linda K. Sussman
1995 The Use of Satellite Imagery and Anthropology to Assess the Causes of Deforestation in Madagascar. *In* Tropical Deforestation: The Human Dimension. Leslie E. Sponsel, Thomas N. Headland, and Robert C. Bailey, eds. Pp. 296–315. New York: Columbia University Press.
Swingle, Charles F.
1929 Across Madagascar by Boat, Auto, Railroad, and Filanzana. National Geographic Magazine 56(2): 179–212.
Sylla, Yvette, and Eugène Mangalaza
1988 L'Image Representative de la Forêt Betsimisaraka, Série Sciences de l'Homme et de la Société. Toamasina: Ministère de la Recherche Scientifique et Technologique pour le Developpement.
Taussig, Michael
1980 The Devil and Commodity Fetishism in South America. Chapel Hill: University of North Carolina Press.
Thomas, Philip
2002 The River, the Road, and the Rural-Urban Divide: A Postcolonial Moral Geography from Southeast Madagascar. American Ethnologist 28(2): 366–391.
Thomas-Slayter, Barbara P., and Dianne E. Rocheleau
1998 Gender, Environment and Development in Kenya: A Grassroots Perspective. Boulder, Colo.: Lynne Rienner Publishers.
Thompson, E. P.
1967 Time, Work Discipline, and Industrial Capitalism. Past and Present 38: 56–97.
1993a [1971] The Moral Economy of the English Crowd in the Eighteenth Century. *In* Customs in Common: Studies in Traditional Popular Culture. Pp. 185–258. New York: New Press.
1993b Customs in Common: Studies in Traditional Popular Culture. New York: New Press.
Thompson, Virginia, and Richard Adloff
1965 The Malagasy Republic: Madagascar Today. Stanford, Calif.: Stanford University Press.
Tilman, David, Robert M. May, Clarence L. Leyman, and Martin A. Nowak
1994 Habitat Destruction and the Extinction Debt. Nature 371: 65–66.
Time
1942 Madagascar: Aepyornis Island. March 23. Vol. 39(12): 36.
Triolet, Elsa
1977 [1965] Le Grand Jamais. Paris: Gallimard.
Tronchon, Jacques
1986 [1974] L'Insurrection Malgache de 1947. Paris: Editions Karthala.
Tsing, Anna Lowenhaupt
2005 Friction: An Ethnography of Global Connection. Princeton, N.J.: Princeton University Press.
Turner, Terence
1979a The Ge and Bororo Societies as Dialectical Systems: A General Model. *In* Dialectical Societies. D. Maybury-Lewis, ed. Pp. 147–178. Cambridge, Mass.: Harvard University Press.

1979b Kinship, Household and Community Structure among the Northern Kayapo. *In* Dialectical Societies. D. Maybury-Lewis, ed. Pp. 179–217. Cambridge, Mass.: Harvard University Press.

2003 The Beautiful and the Common: Gender and Social Hierarchy among the Kayapo. Tipiti: Journal of the Society for the Anthropology of Lowland South America 1(1): 11–26.

2008 Marxian Value Theory: An Anthropological Perspective. Anthropological Theory 8(1): 43–56.

Turner, Victor W.

1967 The Forest of Symbols: Aspects of Ndembu ritual. Ithaca, N.Y.: Cornell University Press.

UNESCO

2000 The Mananara-Nord Biosphere Reserve. http://unesdoc.unesco.org /images/0013/001392/139281eb.pdf, accessed February 1, 2011.

2004 The MAB Program, Mananara Nord Biosphere Reserve: Integrated Project on Conservation and Development. http://www.unesco, org/mab/capacity /Madagascar/project.htm, accessed March 4, 2004.

2006 The Statutory Framework of the World Network of Biosphere Reserves. http://www.unesco.org/mab/doc/statframe.pdf, accessed November 1, 2006.

UNESCO, World Heritage Convention

The Criteria for Selection. http://whc.unesco.org/en/criteria, accessed April 15, 2011.

World Heritage. http://whc.unesco.org/en/about/, accessed April 21, 2011.

UN Integrated Regional Information Networks

2005 Madagascar: Economy Continues to Grow. http://allafrica.com/stories /200506081031, accessed June 15, 2005.

UNWTO

2006 Tourism Market Trends, Annex: International Tourism Receipts by Country of Destination. http://www.unwto.org/facts/eng/pdf/indicators/new /ITR05 jfrica-US$.pdf, accessed July 7, 2007.

Urry, James

1993 Before Social Anthropology: Essays on the History of British Anthropology. New York: Routledge.

Valensky, Chantal

1995 Le Soldat Occulté: Les Malgaches de l'armée française, 1884–1920. Paris: L'Harmattan.

Valmy, Robert

1959 Porteurs et Filanjana. Revue de Madagascar 6: 35–44.

Vanclay, J. K., and J. D. Nichols

2005 What Would a Global Forest Convention Mean for Tropical Forests and for Timber Consumers? Journal of Forestry 103(3): 120–125.

Van Onselen, Charles

1980 Chibaro: African Mine Labour in Southern Rhodesia, 1900–1933. London: Pluto Press.

Vayda, Andrew P., and Bradley B. Walters

1999 Against Political Ecology. Human Ecology 27(1): 1–18.

Walley, Christine J.

2004 Rough Waters: Nature and Development in an East African Marine Park. Princeton, N.J.: Princeton University Press.

Walsh, Andrew
2001 When Origins Matter: The Politics of Commemoration in Northern Madagascar. Ethnohistory 48(1–2): 237–256.
2005 The Obvious Aspects of Ecological Underprivilege in Ankarana, Northern Madagascar. American Anthropologist 107(4): 654–665.
Watts, Michael, and Richard Peet
2004a Liberating Political Ecology. *In* Liberation Ecologies: Environment, Development, Social Movements. 2nd edition. Richard Peet and Michael Watts, eds. Pp. 3–47. London: Routledge.
2004b Preface to the Second Edition. *In* Liberation Ecologies: Environment, Development, Social Movements. 2nd edition. Richard Peet and Michael Watts, eds. Pp. xiii–xviii. London: Routledge.
Weiner, Annette
1992 Inalienable Possessions: The Paradox of Keeping-While-Giving. Berkeley: University of California Press.
West, Paige
2005 Translation, Value, and Space: Theorizing an Ethnographic and Engaged Environmental Anthropology. American Anthropologist 107(4): 632–642.
2006 Conservation Is Our Government Now: The Politics of Ecology in Papua New Guinea. Durham, N.C.: Duke University Press.
2008 Tourism as Science and Science as Tourism: Environment, Society, Self, and Other in Papua New Guinea. Current Anthropology 49(4): 597–626.
West, Paige, and James G. Carrier
2004 Ecotourism and Authenticity: Getting Away from It All? Current Anthropology 45(4): 483–498.
West, Paige, Jim Igoe, and Dan Brockington
2006 Parks and Peoples: The Social Effects of Protected Areas. Annual Review of Anthropology 35(1): 14.1–14.27.
White, Richard
1996 The Organic Machine: The Remaking of the Columbia River. New York: Hill and Wang.
Wilkinson, T.
1869–1870 Journey from Tamatave to the French Island Colony of St. Mary, Madagascar. Proceedings of the Royal Geographical Society of London 14(5): 372–377.
Willems-Braun, Bruce
1997 Buried Epistemologies: The Politics of Nature in (Post) Colonial British Columbia. Annals of the Association of American Geographers 87(1): 3–31.
Williams, Brackette F.
1991 Stains on My Name, War in My Veins: Guyana and the Politics of Cultural Struggle. Durham, N.C.: Duke University Press.
Williams, Michael
2003 Deforesting the Earth: From Prehistory to Global Crisis. Chicago: University of Chicago Press.
Williams, Raymond
1973 The Country and the City. New York: Oxford University Press.
1977 Marxism and Literature. Oxford: Oxford University Press.

Willner, Ann Ruth
1961 Problems of Management and Authority in a Transitional Society: A Case Study of a Javanese Factory. Human Organization 21(1): 133–141.
Wilmé, Lucienne, Derek Schuurman, Porter P. Lowry, and Peter H. Raven
2009 Precious Trees Pay Off—But Who Pays? An Update. December 9. http://www.illegal-logging.info/uploads/PosterrosewoodUpdate09Dec2009.pdf, accessed May 5, 2010.
Wilson, Peter J.
1992 Freedom by a Hair's Breadth: Tsimihety in Madagascar. Ann Arbor: University of Michigan Press.
World Bank
2004 The World Bank Approves Single-Largest Grant Ever for the Environment in Support of Madagascar's Third Environment Program. News Release No:2004/348/AFR. http://web.worldbank.org/WBSITE/EXTERNAL/NEWS/0,,contentMDK:20201239~menuPK:34463~pagePK:34370~piPK:34424~theSitePK:4607,00.html, accessed December 2, 2004.
2005 World Development Indicators Database http://www.worldbank.org/data/countrydata/countrydata.html, accessed June 3, 2005.
World Travel and Tourism Council
2007 Madagascar: Travel and Tourism, Navigating the Path Ahead. London: World Travel and Tourism Council. http://www.wttc.org/site_media/uploads/downloads/madagascar.pdf, accessed December 16, 2011.
World Wide Fund for Nature / Cooperation Suisse
1994 Vintsy: Trimestriel Malgache d'Orientation Ecologique, no. 6 (September): Front cover.
Worster, Donald
1982 Dust Bowl: The Southern Plains in the 1930s. Oxford: Oxford University Press.
Wright, Patricia C.
2008 Decades of Lemur Research and Conservation: The Elwyn Simons Influence. *In* Elwyn Simons: A Search for Origins. John G. Fleagle and Christopher C. Gilbert, eds. Pp. 283–310. New York: Springer.
You, André
1931 Madagascar: Colonie française, 1896–1930. Paris: Société d'Editions.
Zerner, Charles, ed.
2000 People, Plants, and Justice. New York: Columbia University Press.

INDEX

Page numbers in italics indicate illustrations.

GENESE MARIE SODIKOFF is Assistant Professor in the Department of Sociology and Anthropology at Rutgers University, Newark. She is editor of *The Anthropology of Extinction: Essays on Culture and Species Death* (IUP, 2011).